PATHOLOGY OF MODERN INDIAN SCIENCE

Genesis of its eco-System

Rajiva Bhatnagar

Rajiva Bhatnagar: - rbhatcvl@gmail.com
Mob 9425316943
http://rbhatnagar.blogspot.com/
*

Publisher-Kindle Direct Publishing
ISBN: 9781674285238
eBOOK: ASINB082MRFD8
Available from - AMAZON.IN and AMAZON.COM

$ 9.09

Cover Photograph: Credit: https://freedesignfile.com/tag/drowning/
Cover Design: R. Bhatnagar

This book is dedicated to those
who saw
scientific dreams
turn into
nightmares

DECLARATION

The term

SCIENTIST

is used as a gender-neutral entity throughout the book and use of male pronoun is only to maintain the flow of the narrative

About the Author

Author is a scientist retired from the Department of Atomic Energy after about 37 years of research in the area of Laser Technology. Born in 1943 at the city of Mathura in Utter Pradesh, India, he completed his formal education from University of Sagar, Madhya Pradesh, majoring in Physics. In 1967 he joined Bhabha Atomic Research Centre, Bombay after completing one-year induction course. Thereafter he obtained the degree of Doctor of Philosophy from University of Bombay.

The author has interest in science history, poetry and literature. He has published three literary books, a biography of his father and collections of poems in Hindi.

Acknowledgements

It is a difficult task to write an acknowledgement away from the hierarchy of the government scientific institutions where it is clearly defined. Nevertheless, I will try.

First and foremost, I wish to acknowledge constant and irritating discussions on the topic of Indian science with Shri S K Bhattacharjee, a retired scientist himself. He also happened to have graced the portals of the training school of the department of Atomic Energy couple of years before me. He was kind enough to go through several drafts of the book and often came up with unrealistic suggestions that led to counter arguments.

I cannot forget the spirit of enquiry infused in me in my childhood by two most remarkable individuals and scientists, Dr. Indradev and Dr. Rafeeq A Fatehally for which I feel indebted.

I thank Dr. (Mrs.) B. S. Mahajan, Dr. M. S. Bhatia and Dr. P. Venu Babu for painstakingly going through the manuscript.

I wish to acknowledge the support extended by my family, particularly my wife, Dr. M Bhatnagar, during my unsuccessful attempts to become a true scientist.

Prologue

At the end of the last day of my sojourn with scientific research at a premier institute, while leaving, I looked up in the rear-view mirror to find the sun setting on my career. Concurrently a question popped up in my mind. Why am I feeling a sense of disenchantment? Did I fail to put up my best? Why is it that I could not do what I intended? Did the system fail me or was I not up to it?

I have no hesitation in accepting my own mediocre scientific upbringing that left me gasping for breath many times and only my perseverance prevented me from cyanosis. Throughout my career I came across a multitude of students coming out of the same old education system, proceeding to foreign shores and doing well in science while their compatriots, staying back, struggled in various institutes and universities. Even those who responded to the call of the motherland and returned, appeared crestfallen and undistinguished within a short spell of time. This was appalling, and it goaded me to search for reasons with intent none other than to be at peace with myself.

I hope this, small step in redeeming myself, may cajole the intelligentsia, particularly of the scientific community to introspect, debate and rise from the slumber to ameliorate its own environment.

01/12/2019   R. Bhatnagar

Contents

List of Figures

Abbreviations

AEC	Atomic Energy Commission
AECI	Atomic Energy Commission of India
AEE	Atomic Energy Establishment
AEEI	Atomic Energy Establishment of India
AEET	Atomic Energy Establishment, Trombay
AERC	Atomic Research Committee
AMU	Aligarh Muslim University
BARC	Bhabha Atomic Research Centre
BARCOA	BARC Officers Association
BISR	Board of Industrial and Scientific Research
BRAE	Board of Research on Atomic Energy
BSA	Board of Scientific Advise
BSIR	Board of Scientific and Industrial Research
CRG	Chemicals and Reprocessing Group
CSIR	Council for Scientific and Industrial Research
DAE	Department of Atomic Energy
DNA	Deoxyribonucleic acid
DRDO	Defence Research and Development Organisation
DSIR	Department of Scientific and Industrial Research
ETH	Swiss Federal Institute of Technology in Zurich
HARL	High Altitude Research Laboratory
IAC	Indian Advisory Committee
IACS	Indian Association for Cultivation of Science
IAS	Institute for Advanced Studies
IAS	Indian Administrative Service
ICAR	Indian Council of Agricultural Research
ICMR	Indian Council of Medical Research
ICTP	International Centre for Theoretical Physics
IGCAR	Indira Gandhi Centre for Atomic Research
IIRBU	Industrial Intelligence and Research Bureau
IISc	Indian Institute of Science
IIT	Indian institute of Technology
INP	Institute of Nuclear Physics
IMTECH	Institute of Microbial Technology
INSA	Indian National Science Academy
ISCA	Indian Science Congress Association

ISRO	Indian Space Research Organisation
J & K	Jammu and Kashmir
JISA	Journal of the Indian Society for Agricultural Statistics
KC	K S Chandrasekharan
MLS	Mahendra Lal Sirkar
MMG	Materials and Metallurgy Group
NPL	National Physical Laboratory
OSD	Officer on Special Duty
PSLV	Polar Satellite Launch Vehicle
Pt.	Pandit
SCC	Scientific Consultative Committee
TC	Trombay Council
TIFR	Tata Institute of Fundamental Research
TSC	Trombay Scientific Committee
UGC	University Grants Commission
USSR	Union of Soviet Socialist Republic

*

PATHOLOGY OF MODERN INDIAN SCIENCE

Introduction

The sorry state of post-independence Indian science and the glory of Vedic sciences have been a subject of remorse and trumpeting respectively in different quarters of the Indian society. The votaries include eminent personalities, political leaders and uninformed public. It is a recognized fact that science and technology were well advanced in ancient Indian civilization.[1] The origin of place value system and the concept of zero are well publicized and so are the metallurgical achievements showcased by the Asoka's iron pillar. However, this position of scientific and technological eminence was lost, and the early advantages of science were fritted away by the society. This rise and fall of science were not unique to the Indian civilization. Many other ancient civilizations also reached their peak in their scientific endeavours and subsequently declined because of various reasons: social, religious and political.[2] Some of these civilizations re-emerged and attained prominence in modern times and dominated the world politically, culturally and socially. Indian science regrettably was not amongst these. What are the factors responsible for such an unsatisfactory state? Why is our science in such a sorry state? Why do we fare so low in any index that reflects the achievements of science in the world?[3]

Having lost the position of eminence attained in ancient times why is that we are unable to pick up the threads and gain ascendance in modern day science? The questions that need answers are: are we constrained by our social, cultural, religious, and political ethos to recede in science after a promising start before independence? Is the bigotry of prominent personalities of the time responsible for the present situation? We do adore eminent scientists and put them on high pedestal, but do we ever ask inconvenient questions about their actions? How did these eminent scientists and political personalities of the time influence the growth of science and scientific research? Shiv Visvanathan has a good point about the biographies of Indian scientists[4]:-

"One hears the general declaration that science is about objectivity and truth and yet hagiographies are rampant in Indian science………"

This inherent culture of reverence towards icons in social life in general and of Indian science in particular, excludes their critical assessment and shortcomings, leaving a lot unsaid. The scientist-biographers of eminent personalities also tend to hero worship and sweep inconvenient truths under the carpet. Even the autobiographies inch towards becoming hagiographies and become catalogue of the events and personal memorabilia.[5,6] Rarely the eminent scientific personalities, while penning their memoirs, touch upon the problems of scientific research in general and of their efforts to address them in the institutions over which they presided. Even if the problems are aired there is no indication of the efforts made by top level advisors to take up the issues and the response, if any, of the government.[7]

It is necessary to be objective, while discussing the motives and actions of the icons of pre and post-independence Indian science, in order to understand the quagmire in which the young and aspiring Indian scientists presently find themselves in. The role and actions of the well-known political and scientific personalities of the time, who have shaped modern Indian science need critical scrutiny. It is important to understand why, when the scientific research worldwide was increasingly getting dependent on cooperation, in India it was replaced by mutual recriminations and competitive non-cooperation. The subject of the growth and decline of science in India is so wide and multi-dimensional that it will not be possible to cover all aspects in a single book. We will therefore be very selective and brief in our coverage but will try to put forward an approach which to our mind outlines the deficiencies in the growth and nurturing of scientific research in India. We will also try to assess the benevolent contribution of the colonial rule that established some educational institutions and a rudimentary science infrastructure primarily to serve the commercial needs of the Empire [Part I]. This structure was to become an instrument of governance between the Metropolitan (British) science and the peripheral (Indian) science. This inheritance became the precursor of the government science in free India. We will also select few individuals and institutions; the choice arbitrarily decided by the impact they had on post-independence scientific research and education [Part II]. In doing so we will not discuss the scientific contributions of these esteemed individuals, which are undisputedly of very high standard. We will go beyond social niceties and platitudes to critically examine the personal relationships of the leading personalities of the time and their actions to show how their behaviour was anything but detrimental to the making of the eco-system of science after independence. We will discuss the formation and role of two major scientific institutions that largely guided the ethos and discourse on scientific research. In order to identify the factors affecting the growth of post-independence science, we will discuss briefly its growth over centuries of human existence and the emergence of western science in different countries from the cauldron of social, religious and political events and try to establish some parameters against which present eco-system of Indian science can be judged [Part III]. Lastly, we will assess the position where Indian science currently stands and suggest some possible approaches to mitigate the situation [Part IV].

In doing so we will approach, the subject 'With Malice towards One and All' to use a phrase from columnist Khushwant Singh.[8]

*

PART I

The Inheritance

1

Pre-Independence Indian Science, Institutions and Individuals

Western science came to India with the arrival of the European explorers, the Dutch, the Portuguese, the French and the British who came as traders along with Jesuits. The arrivals started the exploration of the newly found territories. With the starting of education for the natives and introduction of medicine and science in the curriculum, the stage was set for natives to get involved with modern science. Modern science came to India with the Europeans in the year 1542, when St. Paul's College and the first printing press were founded by Portugal's Jesuit Francis Xavier in Goa.[9] The Jesuits activities however were banned and were revived only after 200 years, this time by British.[10]

East India Company came with the approval from British parliament to carry out explorations with the aim of opening trade and bring benefits to the company shareholders. It is no wonder that the company came to India to trade and later went on to exploit its natural resources as it gained foothold. India's brush with western science started with the arrival of these explorers who brought in the products of science along with them. Their inquisitiveness and their exposure to untapped natural resources of India turned many of them to amateur scientists. 'In late-eighteenth and early-nineteenth-century England, medical colleges were the only educational establishments to offer some semblance of a scientific education'.[11] These doctors became amateur botanists and zoologists. The army engineers became early geologists, meteorologists and astronomers.[12] Pratik Chakrabarti has briefly discussed the work of scientists in Electro-galvanism, trigonometrical survey, Geology, and instrumentation etc.[13] The company administration often did not take kindly to these amateur scientists and their forays outside the realm of their official responsibilities.[14] The conviction of the racial superiority of the English over the native population led to under evaluation of native's capabilities even though several individuals did contribute scientifically.[15]

The British crown took over the reins of the administration after the uprising of 1857. The new administration organised and institutionalised the gains of the individual scientists. First and foremost, to be organised were Geological, Geographical, Botanical and Zoological surveys. Solar Physics and terrestrial magnetism were brought under the meteorological observatories. The Director-General was acknowledged as 'the Principle Scientific Adviser to the Governor-General- in Council.[16] In 1866 the government, by law, established a public museum incorporating all the branches of natural history but the laws did not refer to research or need of science students.[17] Institutes for teaching and research in agriculture[18], medical[19] and Bacteriology [20]were set up. An excellent account of all the scientific institutes established up to 1947 is given in the articles in the book referred above. Large number of medical students graduated from the colleges and Indian Medical Services came up. It was now time to reorganise other activities under the government which lead to the establishment of scientific cadre. In 1902 Board of Scientific Advise (BSA), consisting of all the heads of departments, was formed. BSA was subservient to the Indian Advisory Committee (IAC) of the Royal Society. IAC was not enthusiastic about basic research in India and wanted it to be left to scientists in Britain. This view was also shared by the Government of India.[21] Science administration in Victorian India was a top-heavy structure, had inner contradiction and professional jealousies.[22] Excessive administrative control exercised at different levels ensured that the colonial scientists always danced to the official tune. This bred a sense of dissatisfaction among the scientist.[23]

Between 1900 to 1920 during the tenure of Lord Curzon a new industrial policy was formulated and Sir Thomas Holland, a geologist played a significant role.[24] He was the first to argue that the interests of science and industry were bound together.[25] Holland sought autonomy for scientific research and revolted against BSA and in the process brought to the fore the differences between the colonial science, administered from London and the scientific objectives of men on the ground. Holland linked science to survival and national security and was one of the most active promotors of the industrialisation of India.[26] Holland's Board of Scientific Advise (BSA), renamed as Department of Scientific and Industrial Research (DSIR) at the onset of World War I provided the structural basis for the Indian counterpart, Board of Industrial and Scientific Research (BISR).[27] BISR, retained by the Imperial government after the war, was subsequently rechristened as the Council for Scientific and Industrial Research (CSIR) to become the backbone of the centralised scientific research in India.

British left India to its own travail in 1947, leaving behind a legacy that was carried forward by the new administration.

The Asiatic Society {(Bombay)-exclusively English until 1840}[28], founded in 1784 gave a common platform to English amateur scientists for publication of their work.[29] Under the influence of the Orientalists, the society's commitment to the Orient restricted the growth of professionalism in science and by the end of nineteenth century the society came to be known only for its cultural studies.[30] Orientalism[31] perhaps negatively influenced the growth of science in a society subjugated for centuries and seeking moorings afresh. It recentralised the discourse on the glory of India's ancient past.

In 1876 Mahendralal Sircar[32] started an organization, Indian Association for Cultivation of Science (IACS) devoted to promotion of science. IACS had a science laboratory and it was hoped that the general public will be educated in science through lectures and demonstrations.[33] Sircar was the first person who brought pure sciences, like physics and chemistry, ignored hitherto by the British, to the fore. After his demise in 1904 his son Amritlal Sircar took over the association and in 1907 he enthusiastically welcomed Raman, a complete stranger to him, to work at the laboratories of the association[34]. At the same time two other philanthropic efforts led to the establishment of the Indian Institute of Science by Tata Trust in 1909 at Bangalore and the Bose Institute by Sir J C Bose in 1917.

Institution building is a complex process and magnificent buildings are only a necessity but not a sufficient condition for a creative and vibrant institution. One can recall the tenacity and persuasion of Sir Asutosh Mookherjee[35] in overcoming the barriers of regionalism to offer C. V. Raman the Palit chair at the Calcutta University.

The institutions develop a character of their own and their ethos are built on the integrity, sincerity and honesty of these pioneers. They shoulder the responsibility to establish, by their deeds, a value system and the ethos that guide the students and teachers alike. These pioneers attract and then nurture students in their formative years to place them on the path of enquiry. M N Saha, S N Bose and D M Bose who made their names in science were the illustrious students of J C Bose and Asutosh Mookherjee. Science, or for that matter any field of creativity, one can say, progresses when an individual becomes an institution himself. Dr. Abdus Salam advocates of building of institutions around these towering individuals, whom he termed as 'the tribal leader'.[36] William A Blanpied identifies six individuals whom he categorizes as 'Tribal Leaders' namely J. C. Bose, P. C. Ray, C. V. Raman, S. N. Bose, M. N. Saha and H. J. Bhabha.[37] While discussing their contributions he concludes that basic research, conducted in a developing country, can contribute to international scientific progress and substantially contribute to the developing country itself. Citing the case of Bose and Raman, he makes a point that the contributions to science from these scientists were exploited by the west.[38] Blanpied, influenced by Bhabha's approach in training manpower for scientific work, gave a new criterion for measuring the scientific capabilities of a country and writes:[39]

"The ability of a less developed country to capture basic research results that originate elsewhere may be a far better measure of its scientific capabilities".

If this axiom is accepted as a measure of the scientific capability of a developing country, it will lead to scientists perpetually wallowing in backwaters of science while the horizons keep receding.

In the Indian subcontinent of British era there were indeed many prominent individuals or the "tribal leaders", who contributed significantly to international science. Of these leaders, the role of M. N. Saha, C. V. Raman, S. S. Bhatnagar and H. J. Bhabha had a large bearing on the ethos of the post-independence scientific community since they deliberated on the problems of science and technology at national level. Did these tribal leaders, who were to become role models for the younger generation of scientists, and to whom the society looked up to for the deliverance, ever pause to ponder the likely effects of their actions on the minds of future Indian scientists who find themselves cursed to wear the albatross of the inherited value system? As we will see the clashes of the egos, maneuvering of the scientific spaces, exploitation of political and social contacts, self- aggrandization and internecine feuds governed the behavior of these prominent personalities. This set norms, a value system and an organizational structure that affected post-independence science rather adversely in a way that made the scientific goals appear less clear than the aims of the consolidation of power and position in institutions that were to be the temples of independent India. Commenting on the golden age of science, art and literature in India, S Chandrasekhar says; [40]

"....it is remarkable thing that in the modern era before 1910 there were no [Indian] scientist of international reputation or standing. Between 1920 and 1925 we had five or six internationally well-known men".

He however averred that it gave a wrong picture about science,

"I certainly had a wrong picture. Ramanujan became famous in four years. Saha's second paper produced the ionization theory attached to his name, Satyendra Nath Bose became associated with Einstein in the second or third paper he ever wrote. And Raman made a discovery and got a Nobel Prize. It gave a very glamorous picture, which was alright for these people, people with considerable standing. Although comparing some of them with other great men of science we know, they may not measure equally. But we must remember that these men came from a surrounding and a background that was devoid of modern science".

Is it that we were lulled by the early success of Indian scientists?

*

2

The Spread of Western Science in India

Spread of western science to the colonies under occupation was modelled by Basalla, who was the first to approach the subject. He proposed a three overlapping-phase model for the spread of Western science in the non-European countries. [Fig. 1][41]

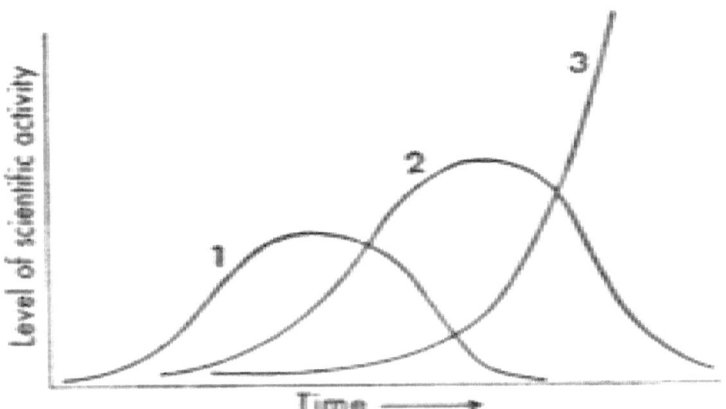

Figure 1 Sequence of phases in the diffusion of Western Science

In the first phase, the colonies become the source for western science. The data is collected in the colonies and analysed at the centre. Basalla terms the second phase as colonial science and the third phase completes the transplantation process. The first and the second phases are intricately linked to the source. The third phase is crucial in establishing an independent scientific tradition in the recipient society. Basalla's model was contested for its drawbacks[42]. It also does not take into account social, cultural, religious, political and economic factors, to name a few, that might influence the spread of science in a society.

Kochhar has interpreted Basalla's model in the Indian context and has designated the first phase as the colonial tool stage where science was used by the British to further their colonial interests particularly in the fields of geology and Botany. [43] Kochhar terms the second stage as the peripheral native stage with extensive involvement of natives.[44] This phase saw the establishment of colleges, setting up of observatories, extensive surveys of the land etc. The third phase saw the coming of age of the native men of intellect. Kochhar names this as 'The Indian Response stage'.[45]

A scientometric study of science in pre-independent India by S. K. Patra and M. Muchie points out that from 1807 to 1858 a total of 99 papers were published. [Fig. 2] [46] The number of publications peaked around 1936 and declined subsequently, regaining somewhat by the time of independence.[47] A total of 20 institutions contributed to the scientific activities. The list of top ten productive authors indicates their primary affiliation to be the universities and colleges. The number of papers published by T. R . Seshadri[48], N. R. Dhar[49] and C. V. Raman[50] were 175, 143 and 74 respectively.[51]

This could be taken as the initial part of the third phase of Basalla's model or Kochhar's Indian response stage. By the time the two phases were completed

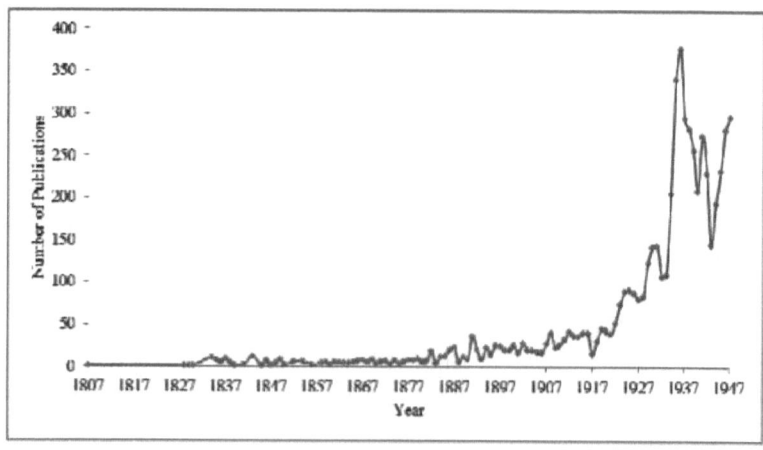

Figure 2 Growth of scientific publications from India during 1807-1947

modern science should have diffused into the culture of the society. In the third stage, science should have taken off on its own, rather exponentially, without reference to the source. Post-independence scientific activity fell in this third stage. It is in this phase that the role of the scientists, scientific societies, general public and the government become important in defining the coefficient of growth. We will, in the coming sections try to enumerate various factors that affected this coefficient.

In 1783 a bill for English education for the natives was introduced in the British parliament.[52] The bill was opposed and defeated on the grounds that the English folly of introducing education in America led to its secession from the British Empire. However, in 1813 the British Parliament passed a resolution stating that

"such measures ought to be adopted, as may tend to the introduction among them [natives of India] of useful knowledge, and of religious and moral improvement".[53]

The emphasis was on religion and moral teachings. Modern science education, that would have offered benefits of Britain's industrial revolution to the Indian subjects was evidently excluded. The importance of education in ushering in the industrial revolution cannot but be overstated. The British rulers did not take adequate measures to help Indians develop science and technology and instead focused more on arts and humanities. The first science subject to be introduced in Hindoo college, Calcutta in 1872 was Chemistry. The University of Bombay offered a B. Sc. degree in sciences followed by Calcutta in 1907.[54]

By late 19th century India had lagged in science and technology and related education. However, the nobility and aristocracy in India largely continued to encourage the development of sciences and technical education, both traditional and western and by 1901 there were 5 universities and 145 colleges, with 18,000 students (predominantly male). The curriculum was Western. By 1922 most schools were under the control of elected provincial authorities, with little role for the national government. In 1922 there were 14 universities and 167 colleges, with 46,000 students. In 1947, 21 universities and 496 colleges were in operation[55]. It is interesting to note that of the total amount spent on education in the 18-year period 1813-1830, as much as 76% was spent in the Bengal Presidency, 19% in Bombay and 5% in Madras[56] It is no wonder that Bengal in general and Calcutta in particular, provided the intellectual backdrop for the cultural, social, scientific and political change. Earliest college to be established was Hindoo College later named as Presidency College, Calcutta (1817). After Macaulay's statement in the British Parliament in 1935 about two dozen colleges were established by 1857 [57] and about six universities were established by 1882.[58]

The industrial revolution in Britain and Europe started in middle of 18th century.[59] The East India Company had gained foothold by that time and the first Science college at Calcutta was established almost a century later. This was also several centuries after the university of Oxford was established.[60] Thus the British knew of the importance of the role education played in modern science and industry and but the advancement of their subjects did not fall within their priorities.

*

3

Provisional Government and the Societies

of Sciences

With the rise of Science education, the number of active workers of science increased. The first and foremost social behavior of scientific workers was the realization of the absence of a common ground for exchange of ideas and some associations were formed. Most associations were established on the lines of their counterparts in the western world. Some of the European scientists, educationalists and philanthropists were at the forefront of establishing these associations. Royal Asiatic Society of Bengal [61] was drawn on the lines of the Royal Society[62] with similar objectives and activities. The Indian Association for Cultivation of Science, also founded on lines similar to the one in UK, sensing a need for scientific research, went beyond and tried to provide a science laboratory where member of the public can engage in scientific research. The Indian Science Congress formed in 1914 had objectives similar to that of the British Association for the Advancement of Science.[63] It is interesting to note that from 1914 to 1930 a single organization served the needs of the scientific community. However, within a short span of 5 years (1930-1935) three more academies sprang up. The objectives of the three were almost similar namely, to advance science in India, publish proceedings, popularize science, hold meetings of scientists, safeguard their interests, render advise to the Government on scientific and other matters etc.[64] The twenty's and thirty's of the twentieth century were the years of profound events on the national scene that were pointing to the impending freedom from the British Rule. These events were probably at the back of the minds of the eminent scientists who were in the process of organizing their activities. It is no wonder that the same period saw an unseemly haste in jockeying for a position of influence in the incoming administration. The Government of India Act was passed in 1935 to grant provincial autonomy. In 1937 provincial election was held and the Congress emerged with majority in seven provinces. The dominant political party, the Indian National Congress and particularly Pandit Nehru was keen to enlist the support of scientists for advising the party. Was the spurt in the efforts towards forming the societies, a result of the anticipation that the days of the British Raj were numbered? Was it an attempt to wrest dominant representational position with the incoming administration?

The importance of science for uplift of the masses and as a means of accelerating the development of the society was realized early by Indian National Congress, the dominant political party. The National Planning Committee (NPC) was setup in 1938 by the then Congress President Netaji Subhash Chandra Bose with M. Visvesvaraya as chairman.[65] However, Saha persuaded Visvesvaraya not to accept the position in favour of Pt. Jawaharlal Nehru. Saha reasoned that without a prominent political leader at its helm the recommendations of the committee may not be taken seriously by the government. The committee composed of scientists, members from industry, eminent personalities from other walks of life produced 27 volumes of reports outlining a comprehensive and integrated plan for independent India's development. These reports however never saw the light of the day.[66] Mahatma Gandhi opposed the very idea of the committee and vetoed the publication of the reports.[67]

In December of 1929 Saha, writing in the Allahabad University magazine, emphasized the purpose of his academy and stated its objective of influencing the government policy.[68]

"the ultimate purpose of the Academy ought to be; to recognize the scientific workers ....and to .... persuade them to take more interest in the scientific matters of national interest; and to exercise a healthy influence on Government in its administrative policy regarding scientific matters".

The first academy of science, the United Provinces Academy of Sciences, Allahabad was thus established. Towards the end of 1930 the Government of India wrote to provincial governments, scientific departments, learned societies, universities and the Indian Science Congress, seeking their opinion on the desirability of forming a National Research Council that would adhere to, and cooperate with the International Research Council and its affiliated unions.[69] The proposal was considered by eminent scientists whose views regarding the composition and functioning of such a National Council were put up in the form of a resolution to the Indian Science Congress Association (ISCA) during its Pune Session.[70] A special meeting of ISCA was held in Mumbai in January 1934 to consider the scheme. In response to the plea made by the President of the ISCA,

Professor M. N. Saha supported an Indian Academy of Sciences on the model of the Royal Society, London. The General Committee of the ISCA unanimously accepted the proposal for the formation of a national scientific society. The Committee formed an 'Academy Committee', which was requested to submit a detailed report for consideration at the next session of the ISCA. [71] G. Govil, writing on the emergence of two academies, states; [72]

> " ....*Delineating the scope and functions of the Academy, which was proposed to be set up in contradistinction, yet complementary, to the aims of the Indian Science Association, Raman wrote; ' The Academy should acquire the necessary authority to advise the Government, the universities and institutions on all scientific matters and other problems referred to it for consideration and to negotiate on behalf of Indian scientific workers with similar institutions abroad' ".*

Raman was unhappy with the representation of south Indian scientist on the drafting committee. Contrary to the avowed pious objectives of making the new academy free of regional chauvinism, to which Raman himself was a votary, he broke away from the deliberations and convened a meeting of the south Indian scientists at Bangalore. The meeting was held under the presidency of C. V. Raman on 1 April 1934 and a society under the name of Indian Academy of Science was formed with Raman as the first president.[73] He remained the president of the academy until his death in 1970.

Saha and others went ahead and formed the National Institute of Sciences of India and its inaugural meet was held at Calcutta on January 7, 1935. Its first president Lewis Leigh Fermor[74], a chemist and geologist, was the Director General of Geological Survey of India. The name of the academy was later changed to Indian National Science Academy (INSA) and its headquarters moved from Calcutta to Delhi.

On the eve of independence, the Scientific Consultative Committee (SCC) was reconstituted and moved to the Department of Industries and Supplies under the charge of C. Rajagopalachari (Rajaji).[75] Shri Rajagopalachari, imploring '*the great scientists of India*' made a serious attempt to unify the three science bodies namely, United Provinces Academy of Science (Allahabad), Indian Academy of Science (Bangalore) and National academy of Science (Delhi).[76]

In the meeting convened by Rajaji, Raman suggested that; [77]

"the Scientific Advisory Committee should be a high powered and authorative body which the Government of India could consult on research matters. The members of SCC should not be chosen by the Government but should be representative of the outstanding scientists of India. That is why a civilized country has a Science Academy. Such an academy must acquire the necessary authority to advise Government".

Mr. C. Rajagopalachari was greatly impressed by Raman's suggestion which tantamounted to an academy of all sciences, combining the existing three academies. Office bearers of the three academies agreed to the proposals and agreed to get the resolutions approved by the respective bodies.[78]

"However, these scientific bodies reneged on the decisions taken in the meeting. Allahabad and Bangalore academies wanted to continue to hold their respective properties. "Raman said that legal merger of properties is impractical".[79]

Raman also had some other reservations. According to Ramasesan, Raman was the member of the founding council of National Institute of Sciences. However, he objected to the presence of several undistinguished members in some academies. He averred how Indian science can prosper under an academy whose first president is a foreigner.[80] Raman was objecting to Lewis Leigh Fermor as president. Fermor was a distinguished geologist in his own right and was the Director General of the Geological survey of India. Raman's objections, though relevant, are unexplainable. In fact, there were many organizations that were started by foreigners on the lines of similar organization in Britain and elsewhere. Raman's own academy had a host of eminent western scientists as fellows' right from its inception. His objection therefore could be construed as putting a spanner in the works. Mr. C. Rajagopalachari observed;[81]

" I find that as things stand nothing further can be done. I do not believe it is at all practicable to get a merger of properties in a legal sense. Unfortunately, this is stated to be an essential condition for the National Institute of science to come in. I would strongly suggest that Bhatnagar and Banerjee persuade their respective bodies to withdraw the conditions as to merger of property and quota of Fellows respectively, so that the United Academy of Sciences of India could be formed, a unity amongst scientists is established and Government can get the best advice in matters relating to science".

Since there was no agreement on the merger and its modalities the nascent Indian government was bereft of the 'independent advice' from the body of Indian scientists. The behaviour of leading scientists of that time exposes the duplicity of thought and action on their part. Raman's enthusiastic support for a single body advising the government during the meeting of SCC and subsequent contrary action is unexplainable.

Regionalism thus became dominant factor and it took a toll on the collective conscience of the scientific workers. Separate existence of these societies was to haunt the Indian government later. Indian scientific community of the time failed in rising to the occasion and lost an opportunity to provide an independent voice on matters of science to the government. Had there been a single scientific body the government might not have sought the help of the British scientists in organising scientific research. This probably would have also deterred individuals from usurping the post-independence scientific agenda.

*

4

Tribal Leaders and Internecine Feuds

In India, the complex social, cultural, regional and family background of an individual has profound bearing on one's development, roots of conflicts and interpersonal relationships. The four protagonists who influenced post-independence Indian science to a large extent, Bhatnagar, Saha, Raman and Bhabha, had very diverse social, economic, regional, and political background.

S. S. Bhatnagar came from a Brahmo samaj family from Punjab. After the untimely death of his father he was brought up by his maternal grandfather. Later he was taken, at the age of thirteen, under the guardianship of his would-be father-in-law who was the headmaster of a school in Lahore. Young Bhatnagar paid for his education through scholarship and by taking tuitions. In 1919 he went to university of London and obtained Doctor of Sciences degree in 1921. He returned to India and became a professor at the Benaras Hindu University.[82]

Saha came from a trader's family of limited resources. After his elementary education he stayed, away from home, at a medical practitioner's house on the condition that he would help in household chores.[83] In his teens Saha protested against Bengal partition.[84] By another account Saha used to go to school barefooted and this was construed as an affront to the visiting Governor.[85] The extent of Saha's association with the Swadeshi movement in Dacca remains unclear. It appears that he knew Pulin Das, the founder of the Dacca Anushilan Samiti, a terrorist organization, which from its very inception emphasized "secrete training of cadres through physical culture and paraphernalia of initiation vows steeped in Hinduism".[86] After he graduated, he was denied permission to appear for a job in the administrative services due to his political belief.[87] Saha indeed suffered because of his political leanings later also as investigations in to his political affiliation resulted in adverse intelligence reports in his selection to the fellowship of the Royal Society.[88]

Raman hailed from a conservative Ayyer family. His father was a teacher attached to SPG College in Trichinopoly.[89] After his graduation, Raman competed for a job with the treasury and was selected. He rose to a high rank in government service. His zeal for science led him to work, after completing his day's official duties in the laboratories of IACS in Calcutta. He left his assured top post in the treasury to work full time in Calcutta University as Palit professor at the instance of Sir Asutosh Mookherjee the vice-chancellor of the university. Not much is known about Raman's political views though he held Gandhi in high esteem.[90] Open expression of one's views on the freedom movement or on Gandhi might not have been very prudent for someone holding a high position in the Imperial service.

Bhabha was born in a rich and happy Parsi family steeped in western culture.[91] After initial schooling in Bombay, Bhabha sailed to England for further studies. He finished his Tripos and worked with eminent scientific personalities of the time before coming to India on vacation. Bhabha was well connected through his industrialist family and his community.

"Homi's father, Jahangir Bhabha, had grown up in Bangalore and was educated at Oxford. He was trained as lawyer in England, and like his father, Hormasji, had started his working life in Maysore, joining the judicial services of the state."[92]

".... Bhabha's uncle had already been a patron of one of its colleges (Cambridge) and helped to finance the engineering department in that university". [93]

Thus, the four Indian scientists who influenced Indian science came from very different family, social, cultural and regional backgrounds. This in turn influenced their interpersonal relations and caused friction. Interpersonal relations of these individuals with the political leaders also had a bearing on the development of science. Initially Saha had easy access to Pt. Nehru, as a member of Science Advisory Committee (SAC). He was outspoken in expressing himself and did not hesitate to air views contrary to the policies of the government. His concerns particularly on atomic energy was not appreciated. His vociferous expression of views, culminated in his relations with Nehru gradually deteriorating to such an extent that he was compelled to write to him in rather strong words about Raman, Bhatnagar, Bhabha and Krishnan. He pointed out that they kept away from the congress when the scientific community supported them and as soon as Nehru got power in 1946 they began to buzz around him.[94] These outpourings seem to be out of deep frustrations in being made a castaway. Abha Sur justifies Saha's role, as his perspective offers a different and critical understanding of the nature of scientific and technological institutions and ideological priorities of the state.[95] Raman, unlike Saha who was denied a government job due to his political beliefs, was in employment with the British government. Saha was not reconciled to life under the British.[96] The two therefore were not on cordial terms.

The animosity between Saha and Raman is well documented.[97] Saha's antipathy towards Raman dates back to the time he was denied a position after Raman left Calcutta.[98] The animosity became so strong that at the time when Saha was seeking funds from Rockefeller Trust for buying a spectrograph for experimentally verifying his theory,[99] *{Note-1}* Raman 'could not rise above his negative feelings towards Saha and gave adverse opinion, saying that he is not an experimentalist and a good organiser.[100] Raman's comments about Saha's experimental capabilities are not above board as the facts are contrary to this.[101] Saha set up an experimental facility for testing his ionisation theory. Abha Sur, while visiting the Allahabad University in 1994 found that Saha's original setup and vacuum furnace was still in use in spectroscopy experiments.[102] Raman also commented that Saha should have sought funds within India. About funds available to Saha less said the better. Even Ashutosh Mukherjee, the vice chancellor of Calcutta University, was handicapped due to resource crunch.[103] Raman and Saha both were scientists working with limited resources and under adverse conditions, but still personal rivalries played their part. The competitive non-cooperation affected Saha's efforts in garnering funds from the west.[104]

A pertinent question to ask is, was Raman not aware of Saha's work on thermal ionization? Was he not aware of its significance and the fact that scientists around the world were trying to seek its experimental verification? Was he not aware of Saha's efforts in trying to get funds for a spectroscope within India? It is improbable. It is not hard to guess that Raman's actions on number of occasions towards Saha were as a result of payback arising out of personal rivalries between the two during their sojourn at Calcutta university. Saha also paid back in equal measure later. Saha was a member of the governing council of the Indian Institute of Sciences, Bangalore where Raman was the first Indian to be appointed as its director. However, Raman faced the prospect of being removed from directorship within three years of joining the Institute. The Irvine committee, appointed to review Raman's tenure, reported that the scientific production had gone up by leaps and bounds but his administration was far from satisfactory.[105] Saha's involvement in this sordid episode can be inferred from the correspondences between him and Syama Prasad Mookerjee;[106]

> *"Rajendra Singh has documented proof that Shyama Prasad Mookerjee, who was on the council of IISc, had a hand in promoting antagonism against Raman........ He cites a letter from Mookerjee to Meghnad Saha where he wrote he had cancelled his trip to England so he could attend the special council meeting that was to look into Raman's directorship tenure. He wrote, "The Irvin Report is now in our hands confidentially.....The report is most favourable. It is a fair and strong document and Raman has been painted in true colour [s].....practically all that we urged in our joint memorandum has been accepted.....Naoroji, Chandavarkar and Ghandy [sic] feel I must not be away at this stage....having worked at it so vigorously for a year and a half.....".*

Uma Parameswaran in the biography of Raman also cites Max Born's letters to Rutherford which shows that the English group also resented an Indian director and Tatas were unwilling to offend their British partners[107].

> *"Max Born's letters to Rutherford soon after he left Bangalore in October 1936, reveal the state of affairs. Born clearly shows that cards were stacked against Raman from the very beginning. He explains this over a span of several letters. 'The English group resented an Indian director' ......The Tatas were ready to drop a brick on Raman rather than offend their British partners, and they entrusted Professor Aston,......with 'the definite mission to clear up the institute.'.....it was evident to me from the beginning that they [the Irvin Committee] had received instructions beforehand..... His enemies the Tatas and the Bengali members of the Council had made up their mind to get rid of Raman. .... Born also speaks of the Irvin Commission as a kangaroo court.... All the dirty affairs were treated in detail, but no voice raised to take in to account the good intentions of Raman or his achievements at the Institute".*

If scientific output had indeed increased during Raman's tenure, his supposed administrative lapses should have been overlooked for his contribution to the advancement of science.

The seeds for such a parochial behaviour were sown way back in 1919 when, after the demise of A N Sircar, Raman was made the secretary of the association and its chief executive. Raman co-opted in its committee of management a large number of members of his choice including his wife Loksundari and brother C Subrahmanyam Ayyar.[108] {Note-2} In May 1933 in a meeting of the management committee of the Association Raman took control of the association and managed to deny election as member to many, including Syama Prasad Mookerjee. His proposal for K. S. Krishnan to be its secretary was accepted.[109] {Note-3}

Raman joined the Indian Institute of science, Bangalore in 1933 but retained his hold on the activities of the association, to the discomfort of his distracters. Raman wanted to retain complete control on the association by changing the process of induction of new members to their fold. One could become member of the association by paying a lump sum to M. L. Sirkar endowment fund without nomination proposal. Raman wanted these to be wetted by the Committee of Management. The proposals were to be placed before the general body of the association. Before the meeting on June 19, 1934, Syama Prasad Mookerjee, with the help of Saha, inducted a large number as members by soliciting contributions to the endowment fund. When Raman arrived at the meeting, he was outnumbered and overwhelmed. [110] {Note-4} Raman resigned from the presidentship and also from the Palit Professorship.[111]

The two linked episodes unfortunately set the agenda for the scientists of the future, namely scheming against colleagues and indulging in groupism. Born's letter to Rutherford exposes the underbelly of the premier institute of the time and shows how the industrialist, Tata, while loosening the purse strings were nevertheless interested in the affairs of the institute and philanthropy and self-interest became coterminous.

What transpired between the two giants of physics over the years that culminated not only in personal rivalries but also of Bengali and south Indian divide? Saha must have been aware of Raman's adverse opinion on him on numerous occasions and must have felt its effect on his scientific work and personal front.

Regionalism has an ugly face that runs as undercurrent in Indian society along with several other divisive attitudes. Raman had earlier faced opposition from the other scientists of the Calcutta university soon after his appointment as Palit professor.[112] Raman's attaining fame in an unfamiliar setting after winning the Nobel Prize must have prompted adversaries to see ghosts where there were probably none. The accusation that south Indians were given preference over Bengalis came into circulation.[113]{Note-5} However, a casual look at the list of Raman's co-authors during his stay at Calcutta reveals that this is far from truth.[114] There are about twelve co-authors from south India as against ten from Bengal until the time Raman left Calcutta. The co-workers from the south, in a completely Bengali milieu, would have definitely become an eyesore. Genealogical data of Raman's co-workers, available on the internet does reveal that some of Raman's students were indeed close relatives. This might have given strength to the accusation of bias against Bengalis. After Raman moved to Bangalore, students from other regions of the country except from south, were not privileged to work with Raman. From 1934 onwards, the list of co-authors reveals that out of seventeen only two were from north and certainly none from Bengal. It might simply be the case of once bitten twice shy. However, this discrimination also is undesirable as it is against the grain of science i.e. being universal and open to all.

Another aspect that comes to fore is that the respect for individual's contribution to science is clouded by extraneous factors like inter-personal equations, regional affiliations etc. Even belonging to the same community does not guarantee respect for one's science. Respect for fellow scientist, even if extreme differences on scientific matters exist, is an essential part of being open to criticism. The controversy between Chandra and Eddington, though bitter, did not cloud the respect and feelings that Chandra had for him.[115] Contrast this with the open hostility between Saha and Raman, where in Saha went on to belittle Raman's discovery in an open forum, in the presence of Somerfield.[116]

In the scientific setup under the government the motives of scientists change. Chandrasekhar has aptly summed up the situation post-independence. [117]

"Motives for which one does science change with time. They should, they must change with time. You see, if I am right in believing that the primary motivation for the flowering of Indian science between, say, 1900 and 1930 was a part of the national consciousness, and that people somehow wanted to show that they were equal to the British, if that was the motivation, it was alright at the beginning and for some length of time. However, when recognition, glamor, reputation continued to dominate as the primary motives of individual scientists' lives, deleterious effects took hold. Those who had made significant contributions were constantly aware of those successes. They wanted to be regarded as unique individuals, and therefore they turned around and discouraged younger people or attributed all kinds of motives to their contemporaries.

5

Breaking of a Bond

The Indian society, from ancient times, has prided in the sanctity of the relationship between a teacher (Guru) and his disciple (Shishya), keeping the Guru on an extremely high pedestal, even above the supreme God. It is pertinent in this light to examine the relationship in the case of Noble laureate Sir C. V. Raman and his industrious student K. S. Krishnan. Krishnan joined Raman in 1923 and was involved in studying the scattering of light, a phenomenon, which later was known as Raman effect. It has been said that Krishnan's contribution to the discovery was no less than that of Raman and yet Krishnan did not get his dues. Sequence of papers published by Raman and Krishnan on the observed phenomenon throws some light on this inglorious episode. S. Chandrasekharan in conversation with Wali, his biographer, says;[118]

"Well, Krishnan did tell me something during that long conversion. If you go back and read the literature, you will find that the first announcement of the discovery was made in a letter published in Nature [31 March 1928], signed jointly by Raman and Krishnan. Then the first spectrum of the Raman effect with its correct explanation appeared later in another joint letter in Nature [5 May 1928]. However, between these two letters, there is a letter signed by Raman alone, in the April 1928 issue of nature. Krishnan told me that Raman had sent this letter which appeared without Krishnan's knowledge and was apologetic about it to him later".

"Krishnan saw the letter for the first time in print and he could not understand why Raman had published the letter. Raman also gave a public lecture [16 March 1928] that was subsequently published. With the exception of this lecture and the one letter in Nature, all other subsequent letters and publications, including the first account of the discovery published in the Indian Journal of Physics [2, no.4 (1928): 399] are under the joint names of Raman and Krishnan. And indeed, Krishnan felt that the announcement in a lecture and the letter Raman wrote under his name were intended to exclude Krishnan's name as a joint author".

Wali further write what, Chandrasekhar, more familiarly known as Chandra, said,[119]

"'Krishnan also told me that January to April 1928 he had kept a diary of all the events that took place during that period. In1971 on my visit to Ahmedabad I found that K R Ramanathan [an early associate of Raman who was working in Raman's laboratory in 1928 when the effect was discovered] was in possession of Krishnan's diary. In an article which Ramanathan had prepared for the issue of Current Science devoted to Raman (Raman had died a few months earlier), he had quoted extensively from Krishnan's diary. The published version of Ramanathan's article contains none of it".

Chandrasekhar confirms that the editor excised it.[120]

Mallik and Chatterjee in the biography of Krishnan present, in detail, the events leading to the discovery and place excerpts from Krishnan's diary which make interesting readings and completely exposes Raman's actions.[121] The first paper on scattering of light was authored jointly by Raman and Krishnan and published in Nature in 1928[122].

"On 8 March Raman sent another letter to Nature under his single authorship titled 'A change of wavelength in scattering'. It was published on 21 April 1928.....The paper closed with the remark;".
"The preliminary visual observations appear to indicate that the position of the modified lines is the same for all substances, though their intensity and that of the continuous spectrum does vary with their chemical nature".
"This was obviously not true. Krishnan saw the paper only when it appeared in print and knew that Raman's last statement was seriously in error. The error had crept in due to Raman's obsession with the idea that the discovery was an optical analogue of the Compton Effect. On 22 March, Raman and Krishnan sent a third letter titled 'The optical analogue of the Compton effect' to Nature, which displayed the Raman spectrum of toluene and gave the correct explanation of the scattering effect. This makes the third paper one of great importance in spite of its slightly misleading title".[123]

The manner in which Raman went about hurrying the publication even before confirmation of the experimental data and its explanation only reinforces that he threw to wind the professional integrity as a scientist while announcing the results.[124]

"Raman's address in Bangalore was written up by him as a full-length paper soon after his return to Calcutta and submitted to the Indian Journal of Physics of which he himself was the editor. ....Without waiting for the formal publication of the journal number and with the authority at his command, he prevailed upon Calcutta University Press to print the paper immediately and arranged for hundreds of the copies of it to be posted all over the world the same day. This was 31 March 1928".

"...........Within a few weeks of Raman's public proclamation of the discovery, Krishnan was able to photograph the anti-Stokes lines in benzene and Raman was very pleased, According to Sukumar Sircar (op.cit.), Raman felt that the discovery of the anti-Stokes lines was as important as the initial discovery of the modified scattering lines at lower frequencies and he told Sircar that Krishnan deserved half the credit for the discovery and that he would share with Krishnan any award that came to him for it".

"On 29 May, L A Ramdas sent a letter to Nature with the title, 'The Raman Effect and the spectrum of the Zodiacal Light', where for the first time the new phenomenon was referred to as Raman effect." The paper appeared in July 1928".

Who could have prompted Ramdas to write a paper prominently attributing the discovery to Raman alone, can only be guessed. Also, Raman's action of publishing the paper as the sole author without even informing his co-worker seem to be nothing but an attempt to undermine Krishnan's role. Within a few weeks a paper referring to the new phenomenon as Raman effect was published in an international journal.[125]

Raman also made efforts to publicise his work transcending the ethics as editor of a journal. He requested A. S. Ganesan to compile a bibliography of papers on Raman effect. When the manuscript was ready, Raman asked his students to communicate a paper to the journal so that the publication of the next number of the journal could be advance. He not only persuaded the superintendent of the press to prepare the galley proofs but also corrected them sitting in the printer's office the same day and ordered 250 copies for distribution. [126] {Note-6}. It is no wonder, if the editor of a journal himself behaves in such unethical manner, that the Indian Journal of Physics ranks so low on the impact factor.[127]

Sequence of events leading to the discovery were recorded by Krishnan himself in his diary. The existence of the diary became known when excerpts from it began appearing in public domain after Raman's death. Did Krishnan have a premonition of the momentous events that he started writing his diary?[128] The diary was taken away by Ramanathan, after Krishnan's death in 1961, from his family. S Chandrasekharan found it with Ramanathan in 1971 and made a copy of Krishnan's diary and deposited it along with his papers with the Royal Society.[129] *{Note-7}* When the diary was returned to Krishnan's family, it contained only sixteen pages. The diary covers the events from February 5 to March 28. The entry on the last day, that had started to describe the actual discovery, ends abruptly in mid-sentence.[130]

> "......*Between 19 to 28 February many studies were carried out. A number of vapors were studied and the influence of the wavelength of incident light on the phenomenon and got surprising results Krishnan notes. At this point the pages of the diary are missing and we are deprived of the description of the subsequent events. When the diary came back to Krishnan's family it contained only sixteen pages, with the last page, wherein he was starting to describe the actual discovery, ending in mid-sentence............*".

The diary shows how Raman was excited about the discovery, how he didn't believe that all liquids show polarized fluorescence, how they missed observing it in last five years. He also thought it to be Kramers-Heisenberg effect and termed it as modified scattering.[131] *{Note-8}* Later when Krishnan demonstrated the effect to him, Raman was so much overjoyed that he told Venkatswaran that the phenomenon will be called Raman Krishnan effect.[132] *{Note-9}* He went on to state that any award that would come to him he would share with Krishnan.[133,] Krishnan's contribution to the discovery can be judged from the fact that between February 1928 to 1929 out of 14 papers written by Raman and Krishnan 8 were with joint authorship and 4 by Krishnan and two by Raman with solo authorship.[134]

Raman was extremely possessive of the discovery and was wary of sharing the credit with anybody else. It was also known to others that Raman was in correspondence with eminent scientists for getting nomination for the Physics Nobel Prize.[135]

While writing a testimonial for Krishnan in 1932 Raman tried to project that if Nobel award was to be made solely on the work done in 1928 instead of the work done since 1921, Krishnan would have come in for a share of the prize.[136] On this Mallik and Chatterjee write.[137]

"Krishnan had started working with Raman in 1923 and a substantial part of the work even prior to 1928 was on the scattering of light in liquids. .... In fact, the feeble fluorescence was very much detected by him in the experiments done in 1924-25. He even found the scattered radiation to be polarised".
"........... Krishnan was present in the laboratory on all the days when the crucial experiments were performed and many of them he had conducted alone, e.g. detection of the polarization of the 'feeble fluorescence' in pure liquids, the first observation of the anti-Stokes lines in benzene, etc."

Raman contradicts himself in his Noble lecture acknowledging that,[138]

"Krishnan observed a similar effect in many other liquids in 1924 and a somewhat more conspicuous phenomenon was observed by me in ice and glass".

Raman's turnaround was therefore an attempt to discredit Krishnan's role in the discovery. Further, in the Nobel lectures he describes the contributions of his other students as if they worked independent of him, but Krishnan assisted him in the discovery[139]

"Krishnan, who very materially assisted me in these investigations, found at the same time the phenomenon could be observed in several organic vapors, and even succeeded in visually detecting the state of polarisation of the modified radiation from them".

Krishnan's relations with his Guru were coming to an end. Krishnan as a true disciple never said a word about the controversy. He walked out of Raman's shadows and charted an independent course and established himself in a different field. He never claimed ownership of the discovery outside of what was so generously conceded by Raman. Thus, ended a fruitful and ancient Guru-Shishya tradition. [140]

"With Krishnan's departure from Calcutta, the scientific collaboration between him and Raman ended. Their paths in research diverged after 1928 and Krishnan never returned to work on Raman scattering. Raman's own

contribution to the further development of the field of Raman scattering is insubstantial".

Raman was not magnanimous even after Krishnan's death. According to Chandrasekhar, Raman said to a Times of India reporter [141]

"Krishnan was the greatest charlatan I have known, and all his life he masqueraded in the cloak of another man's discovery".

If Raman felt so strongly about Krishnan being a charlatan what prevented him from exposing Krishnan before his death? Raman may be a great scientist but failed as a human being and could not rise above self-interest in promoting himself as the sole discoverer.

What impact Raman-Krishnan controversy had on the growth of Indian science and on the morale of future generation of scientists?

Krishnan voluntarily refrained from continuing in the emerging new field of Raman effect, the importance of which might not have been lost to him. With his background in experimental and theoretical physics Indian science in this emerging research would have gained enormously had he continued. This might have resulted in numerous applications that emerged around the world in subsequent years and Indian science would have gained. The insignificant contribution of Raman in his own field after Krishnan parted ways is attributed to his absence. Brand has objectively analysed the achievements and failings of Raman in his Raman's centenary year article.[142] What Indian science lost can be assessed from the spate of activity worldwide in this field soon after the discovery.

Any cooperative creative activity cannot progress unless there is infusion of fresh minds and science is no exception. An apprentice researcher is tutored the tricks of the trade and guided to be observant and truthful to the results of the study. The relationship between the teacher and the taught is of mutual respect and confidence. It is in this light that the unfortunate controversy needs to be addressed and understood. The bond between a teacher and his student ruptured. The post-independence young scientists were no longer prepared to be the Ekalavya[143] (self-learned). With centre of gravity of the scientific research shifting to government institutions with the hierarchical structure of governance and the new entrant becoming gazetted officer, the sanctity of Guru-Shishya relation was thrown overboard. The new officer scientist, more aware of his rights, stood on equal footing with his guru who had probably nothing to part with and had no other means except the annual confidential reports to bank on to control his shishya and enforce a hierarchical knowledge regime.

*

PART II

Post-Independence Eco-System

6

Centralization of Science and rise of Sir S. S. Bhatnagar

The British Government after taking over the reins of administration from the East India Company in 1858 thought of consolidating the scientific research, which hitherto was being carried out under various departments in different Residencies, under a central organisation. Lord Curzon constituted a committee, Board of Scientific Research (BSR) with heads of various departments. The Royal society constituted a committee, Indian Advisory Committee (IAC) to advise the Imperial government on scientific matters. The British government also sought advice on scientific and technical matters from the Royal Society, a non-government organisation comprising of eminent scientists. The same British government, however, sought to appoint a principle scientific Adviser to assist the Governor General on scientific matters pertaining to India instead of seeking opinions of the British scientists engaged in the peripheral science.

After World War II (1939-1945) the Imperial government decided to abolish Industrial Intelligence and Research Bureau (IIRBU). When the proposal to abolish IIRBU came up, the then commerce member of the Viceroy Council, Sir Ramaswamy Mudaliar, proposed that IIRBU may not be abolished but a Board of Scientific and Industrial Research (BSIR) be formed to take its role. In 1940 the Indian government appointed Sir Shanti Swarup Bhatnagar, Head of Chemical Laboratories, Punjab University, Lahore with the designation of Director, Scientific and Industrial Research to spearhead BISR.[144] Sir Ramaswamy Mudaliar became the first chairman of BSIR. Under this board, twenty research committees were formed. Several leading scientists of the time, including C. V. Raman, M. N. Saha, S. K. Mitra, P. C. Mahalanobis, J. C. Shah and S. S. Bhatnagar also found place in some of these committees either as a member or as chairman.[145] This was the first time that eminent Indian scientists were invited to formulate the scientific programme for the country. Each committee met once a year to examine its schemes and plan future programmes. However, these committees represented the second order management level of BSIR, as the final approval of the schemes remained with a body led by largely non-official government scientists.

In 1942 BISR was rechristened as Council of Scientific and Industrial Research (CSIR), an autonomous body, registered under Registration of Societies Act XXI of 1860.[146] Bhatnagar was the architect of the centralization of scientific research. He went on to establish a group of five laboratories aimed at industrial research in various areas in a short time. This was the beginning of CSIR which in due course of time amassed about 38 national laboratories, 39 outreach centres, 3 Innovation Complexes and 5 units employing about 10000 scientists.[147] Bhatnagar's role was not limited only to CSIR. He played an active role in organizing research in nuclear science and some institutions left behind by the outgoing Imperial administration.[148] Notable amongst these were related to Geological survey, botanical gardens, agricultural research, observatories and weapons inspection depos. Defence Research and Development Organisation had its humble beginnings out of the inspection depos of the Imperial era.[149] This grew into a megalithic organisation comparable to civilian science organisation like CSIR and Department of Atomic Energy (DAE) employing thousands of scientists. DRDO, started with about ten institutes and laboratories, now has about 5000 scientists in about fifty institutes.[150] Excellent account of the growth of DRDO is given by Shenoy.[151] DAE has about 30 institutes and organisation encompassing all nuclear related activities, employing about sixty thousand personnel. The organizational structure of the new institutes is hierarchical except the Defence laboratories that have duel controlling authority of civilian and military officers.[152] The structure however is also hierarchical as that in any other government organization.

The centralization and growth of scientific institutions, in post-independence India, is synonymous with the growth in power and prestige of S. S. Bhatnagar and H. J. Bhabha, who set the agenda for post-independence scientific research. Sir Shanti Swarup Bhatnagar is credited with setting up of several research laboratories under the CSIR umbrella. Bhabha, on the other hand, was instrumental in putting India on the international atomic energy map through establishment of TIFR and then expansion of activities under the DAE. While CSIR was the baggage carried over from the colonial times, the Tata Institute of Fundamental Research (TIFR) and DAE were minted afresh with DAE taking shape completely after independence. CSIR and DAE became synonyms, respectively, for their founding fathers, Sir S. S. Bhatnagar and Homi J. Bhabha. The actions of the two individuals had a large bearing on the norms of behavior of the government scientists.

Prof. S. S. Bhatnagar became the first Indian scientist to head a government scientific organization. He was also the first, in the line of several distinguished university academics, to be moulded into a government official with all the magnificent trappings of the position of power. Anderson, captures the essence of the transformation of a university academic into a government scientist with all the trappings of the Raj,[153]

"Shanti Swarup Bhatnagar arrived in Delhi at the beginning of the hot summer of 1940 and realized that there was nowhere to put his Steel Scholars and nowhere to build his promised research labs. ....By custom, the whole government of India moved to Simla in April and stayed three months up in the cool hills. As the advisor and soon-to-be director of scientific and industrial research, Bhatnagar joined this annual escape from the heat. In this intimate atmosphere of a small hill town, he made high level contact essential for his career and his new organization".

In 1943, Prof. A V Hill was invited by the Governor General of India to report on the state of scientific research in India. Hill toured the length and breadth of the country and submitted his report on 14th August 1944.[154] Hill recommended the formation of six boards to encompass all the scientific activities scattered under various government departments. He recommended funding these scientific activities under a central organization. Deepak Kumar points out that;[155]

"Hill himself argued for centralization (which he would not prescribe for Britain). Centralization and concentration of power were to become hallmark of the scientific establishments in post-independence India."

Pt. Nehru, the Prime Minister of the first government of free India was deeply influenced by the progress made by Soviet Union. He opted for a mixed economy with state control in major sectors of industry and private enterprise in some. Nehru was probably influenced by left leaning Noble laureates P. M. S. Blackett[156] and J D. Bernal.[157]

In 1947 Patrick Blackett [158] was asked, during a lunch in Nehru's house in Delhi, on ways to Indianize the military.[159] Blackett's opinion resonated with Nehru and he went on to shape not only the preparedness of the Indian military, but also the civilian Indian science. Blackett's proximity to the first Prime Minister of Independent India, and close interaction with the scientist-administrators like S. S. Bhatnagar, H. J. Bhabha, D. S. Kothari and others is discussed by R. S. Anderson in detail.[160] The profound influence Blackett had on the establishment of scientific research can be summed up by the fact that his association with the Indian Prime ministers lasted from 1947 until his death in 1973. He visited India on innumerable occasions and met with a galaxy of powerful Indian scientists and military generals, visited almost all the important institutions including those covered by stringent confidentiality laws.[161] Anderson, in his two part article, has also brought out influence of Blackett in defence and scientific matters.[162] Blackett worked as adviser to the prime minister without holding any position in the government. As an adviser he had no accountability. His easy access to the prime minister placed him in a unique position that made him a conduit for the Indian scientists to have accesses to the prime minister. Blackett himself gives more importance to his contribution as military consultant rather than his role as scientific inventor. Indians however perceived his importance for his influence on Indian science.[163] *{Note-10}*

Delivering the Jawahar Lal Nehru Memorial lectures in 1972, Prof. Blackett recounts his visits, on several occasions, to India and his pleasant stay at the PM house and truthfully describes his role in advising the government not only on defence matters but also on problems of civil science and education.[164] Why was he so influential? The western education and ethos became a common ground for intellectual discourse between these advisers, Indian scientists and the political class to the exclusion and disadvantage of the rustic scientists like Saha and Raman. Blackett's another Cambridge connection was H. J. Bhabha whom he knew from 1930's and had personal relations.[165] Another friendship that mattered to Indian science was between Blackett and Sir S. S. Bhatnagar.[166] The other scientists with whom Blackett was in personal contacts were D. S. Kothari and P. C. Mahalanobis.[167] Anderson writes about Blackett's influence on science administration through the Indian scientists. The Indian scientists travelled abroad regularly, to keep contacts with their students, select candidates for appointments and asked Blackett and Hill to watch out for theirs wards and relatives when studying or working there. They were also friends with Bhatnagar and Bhabha[168] *{Note-11}* Blackett's influence in higher echelons of power was so much that he wrote appreciatingly to the Defence Minister when Dr Kothari was appointed to head the Defence Research and Development organisation (DRDO).[169] His influence in government circles was so much that he was channelling all requests for employment to Kothari[170] *{Note-12}*

Blackett's influence in scientific, political and defence matters was profound and was also instrumental in soliciting support for Bhabha. Anderson writes[171]

> *".... Despite the chaotic changes going on around them, his friends in India correctly evaluated his value to their objectives. In the period between 1945 and independence, Britain was preoccupied with many issues in India, but scientific development was not really among them. As the end drew near, after his intensive meetings in India with members of the Atomic Energy Committee (Bhabha, Bhatnagar, Meghnad Saha) and Nehru in early 1947, Patrick Blackett briefed the Viceroy and Field-Marshall Auchinleck in Delhi, and Prime Minister Atlee, Sir Stafford Cripps and Lord Mountbatten (poised to be appointed to be the last Viceroy) in London. The subject of all these meetings was 'the atomic energy set-up in India', and it appears that he was also basically soliciting their commitment to assist his friend Homi Bhabha, whom Nehru had already identified as his champion of atomic energy and nuclear research".*

Heads of Indian institutions looked up to Prof. Blackett, to solve their administrative problems. Blackett was asked, by the minister of Scientific and Cultural affairs, after the death of K. S. Krishnan, to address the problems in CSIR.[172] Blackett prepared his report and recommended moving most of those working in the basic sciences, radio, glass and ceramics out to industries and suggested reorganisation of remaining divisions to enhance communication amongst them. He also commented on the administration;[173]

> *"Blackett found a culture of administrative rigidity in the NPL, where most people fought to define and protect the boundaries of their work. He thought this could be overcome by reorganization. In effect, he said, the NPL lacked purpose".*

In 1964 Dr Hussein Zaheer, the Director General of CSIR was apprising Prof. Blackett about the removal of the Director of NPL who was brought in to implement the changes recommended by him.[174] It is beyond comprehension that these eminent scientists were at sea dealing with interpersonal administrative problems.

Saha was opposed to the interference of the European scientists in the internal affairs arguing that no independent country does this.[175]

"I consider that the practice of calling a few European scientists and getting their opinion on items about which only one point is put before them is a very unwise and unstatesman-likc procedure and should be discarded. Let them come to help our workers in our scientific up-building, but they should not be asked to meddle in our internal affairs. I do not know of any independent country which does it".

Saha's voice was not heard.

S. S. Bhatnagar's approach in dealing with the hierarchy of the government was typically a leftover of the Raj. Anderson writes that while visiting the minister in charge, Maulana Azad, Bhatnagar carried with him a bottle of good whisky though he himself never drank alcohol. [176]

"he was a lifelong vegetarian and never drank alcohol, though it is said that when he went to get Maulana Azad's signature on a file in the evening, he would take along a bottle of good whiskey and they talked more about poetry than administration".

This form of paying respect through gifts, while visiting a ruler, so common during the reigns of Kings and Emperors continued during British times that followed it diligently to curry favours from the rulers. Maulana Azad, while paying tribute to S. S. Bhatnagar after his demise recalled meeting Bhatnagar first time in 1942 at Howrah and appreciated Bhatnagar's action of touching his feet in front of thousands of onlookers.[177]

"I met Dr. Bhatnagar for the first time in 1942. I was returning to Calcutta after release from the Central Jail at Naini. As was the convention in those days, there was a crowd of some thousands at Howarah to receive me. When I came out of the crowd and got into my car, one man detached himself from the crowd and came up to me. He touched my feet and said, "I am Shanti Swarup Bhatnagar". The Govt. of India had, at that time appointed Dr. Bhatnagar as Director, Council of Scientific and Industrial Research. As such, he was a

Government servant while I was Congress President and a rebel against the Government. He did not have a moment's hesitation to express his regard, in this manner, for a rebel against the Government in the presence of thousands of spectators and the full complement of the CID staff. During the last 7 years, ...In 1948, he was for a few months, Secretary to the Ministry of Education; I had still greater opportunity of seeing for myself his work since 1952 when I assumed the charge of the Minister of Natural Resources and Scientific Research and Dr. Bhatnagar was, for over a year, Secretary of both the Ministries, He was at that time carrying the burden of four men. He was Secretary to two Ministries, director of the Council of Scientific and Industrial Research and also Secretary of the Atomic Energy Commission...... ".

Bhatnagar was at that time Director of CSIR. He later went on to work in the Ministry of Education under Maulana Azad where he was appointed the Secretary, first non-career civil servant, without approval from the sub-committee and much against the procedures set up by Sardar Patel.[178] *{Note-13}* Why would Maulana Azad go against approved procedure and draw Patel's wrath and appoint Bhatnagar as secretary to his ministry? The answer cannot but be seen as quid-pro-quo. Was Bhatnagar not instrumental in setting up of a precedent for the relation between a scientist and his political masters? Was this sycophancy not setting the norms of behaviour? Expressing reverence to elders and higher up in the power structure by touching their feet might be a social norm, but he should have been careful, knowing that he was setting a precedent.

Bhatnagar also cultivated closeness with the British advisers to Pt Nehru to take advantage in his career. His open expression of near family relationship with Hill and subsequently using this for getting advantage in career cannot be easily understood. Writing to Hill he emphasizes on the close family relationship.[179]

"....as part and parcel of my house....... My wife and children look upon you as if you are a near relative"

What does one make out of such an expression of family feeling towards a person close to political power? Why was it necessary for Sir Bhatnagar to indicate closeness to Hill unless it is for exploiting it? Hill's influence in matters of appointments and promotions was also substantial which can be seen from Bhatnagar's efforts. Anderson writes [180]

""Bhatnagar wanted to be promoted from Director to a position of greater influence in the bureaucracy, saying to Hill, ' I hope you continue to be of the same opinion as before that if they don't give me the position of at least an Additional Secretary I should resign" He got the position and did not need to resign".

Bhatnagar's efforts at cultivating relationships were typical and go well beyond the professional conduct. Making high level contacts continued to remain important, after independence, as a legacy of the Raj and continued to influence development of Indian science. As the first Indian scientist/administrator to the government, he had far greater responsibility to set norms for the coming generations of scientists waiting at the dawn of independence and beyond. What norms did he set; that, apart from personal capabilities, it pays to manage one's contacts and acquaintances with the political power for advancement of one's career.

*

7

H J Bhabha: The Indian Prometheus

Let us begin this chapter by invoking the memory and teachings of Mahatma Gandhi. We recall his musings on the means that one uses to attain the ends. Writing in Hind Swaraj Gandhi says; [181]

"The means may be likened to a seed, the end to a tree, and there is just the same inviolable connection between the means and the end as there is between the seed and the tree."

Why invoke Gandhi in the mundane process of writing about an individual who has been anointed as an Indian Prometheus,[182] a Greek God, by his community.

"In Greek mythology, Prometheus is a Titan, culture hero, and trickster figure who is credited with the creation of man from clay, and who defies the gods by stealing fire and giving it to humanity, an act that enabled progress and civilisation. Prometheus is known for his intelligence and as a champion of mankind and also seen as the author of the human arts and sciences generally."[183]

A person with such extraordinary qualities definitely left unknowingly an impression, by his actions, on the institutions that he created. Bhabha singlehandedly desired and got his dream institution by diligently working towards it. This is in sharp contrast to Sir Bhatnagar building CSIR institutions out of an inherited structure. Bhatnagar did not dream but was thrust upon the responsibility to build institutions.

Tata Institute of fundamental research was the seed which grew into the banyan tree of the department of atomic energy that now employs about sixty thousand personnel. It was a massive effort and Bhabha certainly deserves the credit. However, the events that led to the formation of the institute, actions of the then personalities including that of the founding father carried in its DNA, a system of values and ethics that in the final analysis became the bane of the institutions. It is in this light that we should examine the events leading to the seeding and then go on to examine the growth of the tree. The roots carried the ethos evolved through actions of the founding father.

7.1 The Seed

Unlike CSIR, the Department of Atomic Energy (DAE), started at the initiative of Homi Jahangir Bhabha, a cosmic ray theoretical physicist, was unencumbered with the baggage of the colonial past. The seed of this mega organisation was the Tata Institute of Fundamental research (TIFR), a private institution started by the Tata Trust at the instance of Bhabha. It was to be the first scientific research institute of independent India that was supposed to become the beacon for the scientific research. Study of the process of establishing the institute and along with it the rise of Bhabha on the Indian scientific horizon sheds some interesting light on how an intelligent and ambitious individual, with appropriate connections could manoeuvre and outsmart contemporary scientists in achieving position of power and fame. That seed grew into a banyan tree of nuclear research and technology, dazzling the benign people with its promise of charm and power.

Looking at the history of nuclear research in India, in the colonial times the importance of nuclear energy must have sufficiently dawned upon the physicists of that time so that they took initial steps in planning research in this area. Among the very first Indian physicists to engage with radioactivity research were Ruchi Ram Sahni with Rutherford at the Manchester University (1912-1914), Satyendranath Bose with Marie Curie in Paris (1924-25), Rajendra Lal De with Marie Curie and with Otto Hahn, Debendra Mohan Bose first at the Cavendish, and later with Erich Regener in Berlin (1919).[184] In 1938 Saha and Raman deputed B. D. Nagchaudhury and R S Krishnan to Radiation Laboratory and Cavendish labs respectively for work in nuclear physics.[185] D. M. Bose established a laboratory as far back as 1923, built Wilson cloud chamber with the expertise gained from working with inventor Wilson, experimented with photographic emulsion to routinely study cosmic rays.[186] The centre of gravity of nuclear research then was at the Calcutta University.[187]

Bhabha's appearance on the scene was due to events beyond his control. He was on vacation in India when the World War II erupted, and he found himself exiled from Cavindish.[188] Seeing no end of the war in sight Sir Dorab Tata Trust created a position for him at the Indian Institute of Science, Bangalore.[189]

"He initially stayed on with family in Bangalore, where his father worked in the public instruction office of the Maharaja of Mysore and was a member on the council of the IISc. In November 1939, Bhabha was appointed special reader at Raman's department of physics and by January 1940, he began teaching. Five months later, the Dorab Tata Trusts gave a grant to support Bhabha's experimental cosmic ray physics research".

In about two years' time Bhabha was appointed as professor after he was honoured as Fellow of Royal Society (FRS).[190] Jahanvi Phalkey contrasts the facilities offered to him as compared to an FRS in England;[191]

"He was allowed to employ an experimental physicist and 4 post-graduate students. In a remarkably short period of time, Bhabha had derived tremendous benefit for his scientific talent from his family connection and his location in India. Bhabha had been awarded distinctions prior to his arrival in India, but an FRS would not have necessarily brought a professorship and his own research group if he had continued to work in England".

In Bangalore even while working in the best institution at the time he was socially and culturally cut off. His isolation is aptly reflected in his letter to Pauli (dt. 04/06/43).[192] Bhabha hoped that the war would be soon over and all can turn to purely scientific activity.[193] His letter to Chandrasekhar (Dt. 20/04/44) indicates that he was yearning for a place like Cambridge that has an atmosphere which no place in India had.[194] Bhabha was still not reconciled to stay in India.[195] In his publications he gave his affiliation as "*at present at the Indian Institute of Science*" even as late as 1941, he wrote to Blackett and looked forward to return to England.[196]

It was not that Bhabha lacked any offer for suitable openings and avenues for research. He, according to his own admission, was offered positions, at very favourable terms, at the Royal Institute of science, Bombay, University of

Allahabad and the Indian Association for the cultivation of science, Calcutta.[197] However he did not find these suitable. In a letter to Krishnan he expressed his dislike for teaching stating that he wanted to devote his time to research and not to be bogged down by routine teaching.[198]

The Parsi community, to which Bhabha belonged, was very proud of his achievements and was eager to see him settled.[199] It is in this light that the events leading to the establishment of the institutes devoted to fundamental research and atomic energy should be seen.

As we have seen the pre-independence Indian scientific community was alive to the new field of nuclear physics and were aware of the research elsewhere in the western world much before emergence of Bhabha in Indian science. Nuclear research was started at a number of universities. Saha was the first Indian scientist to propose use of nuclear energy.[200]

"Meghnad Saha had already discussed the issue of industrialisation outside the NPC in 1938. For example, as president of the National Academy of Sciences in Allahabad, he chaired a session on power supply problems where he proposed the use of power from nuclear reactors in 1939. Nehru whose home was in Allahabad, presided.....".

In 1939 Saha was the member of the National Planning Committee (NPC) of the Indian National Congress. DeVorkin writes about Saha's role;[201]

".... Jawaharlal Nehru, the chairman of the planning committee, called for a comprehensive blueprint for India's industrialisation, and Saha made sure it included nuclear power....".

It is clear that committee members had taken stock of the nuclear power in their deliberations. In 1944 the industrialist members of NPC broke away and proposed a plan for industrialisation known as 'Bombay Plan'. It is inconceivable that prior to 1944, when Bhabha wrote that famous letter to Tata Trust, he and the members of the trust would not have been aware of the developments taking place in the government, particularly those related to science and technology considering the fact that Sir Ardeshir Dalal of the Department of Planning and Development, was an associate of the Tatas;[202]

"Dalal was head of the Planning and Development Department in Delhi and a key Bombay leader close to the Tatas".

Anderson rightly poses a question in the light of the priorities facing the government;[203]

"Even in 1947, the numerous other priorities that awaited decisions, sometimes for a long time, made nuclear development a long shot. So how could this improbable thing happen at all? How did Bhabha's unconventional biography change the odds in favour of a national institute for fundamental research in physics and mathematics?"

Answering the question, Andersons writes,[204]

"Positive signals for the planning and support of research institutions were communicated from the viceroy's office even before Bhabha received his FRS in 1944 in Delhi. These signals coincided with Bhabha's realisation that he ought to try to do something creative in India and not leave when the war was over. Bhabha was now in the elite information network through which he would have known all about such high-level signals".

Zia Mian writes;[205]

"In effect Bhabha saw there was a new community of scientists as advisors to governments, men who were as he had put it in his letter to Tata "outstanding pure research workers" and able to " act on the directing boards in an advisory capacity." They were individually shaping the future like none of them had believed possible before".

Bhabha's close confidant Rustum Choksi, a member of the Tata Trust, guided Bhabha at every stage in drafting the proposals that culminated in that often-cited letter to the Tata Trust.[206] The record of the correspondence between him and Bhabha indicates the extent of backdoor consultations to rehabilitate Bhabha.[207]

"...Turning to lesser things, I think you should now submit a scheme to the Trustees. Send it along with a letter to Sir S. D. S in which you may suggest it is for his consideration, that you'd have liked to put it to him in person, and that I'd written to you to send on the scheme. You'd be glad of any suggestions,

modifications. Pile it on a bit! If he felt it could be submitted to his colleagues, would he do so, please? Point out it is an expression of work in hand; suggest Trustees need not commit themselves for more than 5 years. Hint that it may be later joint effort of Trust, Board of S and IR and Univ. Your letter should be personal. Address the scheme to the Chairman, Sir Dorab Tata Trust. Better still, though it may waste time send the scheme as a draft only and say that you will send it formally to him when you've received his opinion".

The persons sounded in this venture were Choksi's brother J. D. Choksi, Bhabha's younger brother, J. J. Bhabha, J. R. D. Tata, Ardeshir Dalal, J. Mathai and Ratan Tata, all influential people of the trust.[208] The culmination of this probably prompted Choksi to enthusiastically start one of his letter to Bhabha with the statement;[209]

"The bread, as you know by now, was well and truly delivered".

What is this bread that Choksi was referring to? Choksi went on to advise Bhabha on the positions and the financial details and also the manner in which the scheme should be submitted. Encouraged by Choksi, Bhabha went on to place the proposals in front of Saklatvala expressing his duty to serve the country.[210]

"But in last two years I have come more and more to the view that provided proper appreciation and financial support are forthcoming it is one's duty to stay in one's own country and build up schools comparable with those that other countries are fortunate in possessing".

This change of heart probably came about after he was assured of the 'Bread'.

Bhabha's inclinations and his efforts to find a job in any western country must have been known to his well-wishers, otherwise even as late as 1944 Choksi would not have had to advise him to emphatically state about staying in India.[211]

On March 12, 1944 Bhabha wrote that famous letter to Sir Sorab D. Saklatvala of Tata Trust that became the seed of an institute the Tata Institute of Fundamental Research. [212] *{Note-14}*

Bhabha in his letter wrote;

".....There is at the moment in India no big school of research in the fundamental problems of physics, both theoretical and experimental. There are however scattered all over India competent workers who are not doing as good work as they would do if brought together in one place under proper direction. It is absolutely in the interest of India to have vigorous school of research in fundamental physics, for such a school forms the spearhead of research not only in less advanced branches of physics but also in problems of immediate practical application in industry. If much of the applied research done in India today is disappointing or of very inferior quality, it is entirely due to the absence of a sufficient number of outstanding pure research workers who would set the standard of good research............ ".

Was this statement based on facts? Bhabha outlined the proposed areas of work for his institute:

".... The subjects on which research and advanced teaching would be done would be theoretical physics, especially on fundamental problems and with special reference to cosmic rays and nuclear physics, and experimental research on cosmic rays............ ".

The proposals bemoan the lack of big school, completely ignoring many scientists and their work. What were the likes of J. C. Bose, Meghnad Saha, S. N. Bose, D. M. Bose, C. V, Raman and hosts of other scientists, in their respective fields doing? How was their work disappointing or of inferior quality? Was their work not of fundamental nature? Saha ionization, Bose-Einstein condensation, mm wave generation of J. C. Bose, Raman Effect and Mahalanobis distance were well known and well established. Raman's work had won him the Noble prize. Why did Bhabha downplay these achievements of fellow compatriots? It is possible that Bhabha was looking at the work of others from a narrow perspective of cosmic ray physics. If so, even in this area he ignored the work of. Prof P. S. Gill and D. M. Bose.[213] P. S. Gill was Bhabha's contemporary who also worked briefly at the TIFR. D. M. Bose was a well-established scientist in the areas of cosmic rays and Nuclear physics.[214]

The work of D. M. Bose is discussed in a number of articles.[215,216,217] It is said that Bose missed the Nobel prize on his discovery of mu-meson,[218] a discovery for which the Nobel prize was awarded to C. F. Powell[219] in 1950. Powell was gracious enough to acknowledge the contribution of Bose and Chaudhuri and wrote;[220]

"In 1941, Bose and Chaudhuri had pointed it out that it is possible, in principle, to distinguish between the tracks of protons and mesons in an emulsion. The method was based on the difference for a given value of residual range, in the momenta of particles of different mass. This has the consequence that the 'scattering' of the particles will be different: the smaller its mass the more the track of a particle deviates from a straight line as it approaches the end of its range. Bose and Chaudhuri exposed 'half-tone' plates at mountain altitudes and examined the scattering of the resulting tracks. They concluded that many of the charged particles arrested in their plates were lighter than protons, their mean mass being 200m$_e$ .... the physical basis of their method was correct, and their work represent the first approach to the scattering method of charged particles by observation of their tracks in emulsion".

Roy and Rajender Singh further write-[221]

"From the above it is clear that D M Bose and B Chaudhury were indeed the first persons who observed the meson tracks in photographic plates. Not only that they have measured the mass of this particle for the first time long before Powell and the measured mass of (~200m$_e$) is quite close to the accepted value (216m$_e$) as was measured by Powell using improved 'full tone' plates."

At the time of Bhabha's writing of his proposals to the Tata trust in 1944, D. M. Bose had published six papers on cosmic rays alone of which three were published in Nature. Bose had established a method for detecting heavy particle in photographic plates and estimated the mass of mu meson from the record of particle tracks. Some of his works like Bose Effect, Bose Theory, Bose-Stoner Theory or Welo-Bose Rule are named after him.[222] Still Bhabha had the temerity to write,

"There is at the moment in India no big school of research in the fundamental problems of physics, both theoretical and experimental". [223]

It is highly unlikely that he was not familiar with Bose's work and Saha's involvement with nuclear physics at Calcutta. Saha and Bhabha had together attended the International conference on nuclear physics, Copenhagen in 1936.[224] Also M. S. Sinha, One of D. M. Bose's student was working under Dr Bhabha at Bangalore on mu-meson scattering.[225] However, he chose to ignore their existence.

Bhabha proposed theoretical and experimental research in the area of cosmic rays for his institute;[226]

"The study of cosmic radiation forms the main field of experimental research at this institute, though I hope that in the near future experimental work will also extend to nuclear physics. The two branches are very closely knit namely the. origin of nuclear forces owes its existence to the discovery of the meson in cosmic radiation. Bhabha found a linking hook where his own interest in particles and cosmic ray research, the agenda of carrying out fundamental research in the new institute and the referent of nuclear physics converged: in the search for the meson".

Saha was not convinced that cosmic ray physics is atomic physics and artificial creation of meson: Saha wrote to Bhatnagar; [227]

"I feel that the programme submitted by Dr. Bhabha is amateurish and fragmentary. What is proposed is not atomic physics but cosmic rays. .... Even if creation of artificial mesons was an interesting problem, ... 'there seems to be no close link of either cosmic rays or very high-energy particles with fission phenomena".

From the documents in public domain, it can be inferred that Bhabha's writing the proposal to Tata Trust, granting of funds by the trust, promised support from the governments, in a carefully orchestrated effort to rehabilitate him was a foregone conclusion. The members of the trust and their contacts in the government saw an opportunity in the twilight era of the British Raj to establish Bhabha.

Tata Institute of Fundamental Research (TIFR) was established in 1945. The institute started in the premises of IISC, Bangalore. Bhabha became its first Director. The institute was moved to Bombay, Bhabha's hometown, within six months of its founding. It was housed in a bungalow where Bhabha was born.[228]

It is clear that Tata trust was interested in settling of one of the family members rather than help Indian science. J. R. D. Tata responding to Dhirendra Sharma's request for a meeting, in a telegram dt 31/3/1981, accepts his role in supporting his nephew,[229]

"I have played no significant role except supporting Dr. Homi Bhabha in early years and serving as a member of AEC which I have only just re-joined. No useful purpose would therefore be served by meeting me".

Is this really a modest submission on the part of J. R. D. Tata (Jeh)? M. G. K. Menon, in a lecture delivered at TIFR says; [230]

"Once the programme got going Homi had Jeh as a Member of the AEC, in 1962. Jeh continued to be a Member of the AEC even after Homi's death. Jeh was a Member of the AEC for 26 years in all.

Bhabha, in a note regarding organisation of atomic research in India submitted to the then Prime Minister, had invoked secrecy and written;[231]

"The present Board of Research on Atomic Energy cannot be entrusted with this work since it is an advisory body which reports to the Governing Body of the Council of Scientific and Industrial Research, composed of 28 members including officials, scientists and industrialists. Secret matters cannot be dealt with under this organisation."

In essence Bhabha wanted an organisation independent of any other outfit and to be answerable directly to the Prime Minister. How could he have ignored the confidentiality clauses and justify the presence of non-government individuals as members of the commission? If industrialists were to be on board for some reason, then why not others? These are the disturbing questions which point to the usurping of government's scientific agenda.

Bhabha proposed and got an organisation whose activities were supposed to be shrouded by strict secrecy laws and to which very few scientists were privy to. From 1948 to 1958 all members of the commission were scientists. The participation of scientists declined to two out of three from 1958 to 1962, one out of four from 1962 to 1965, one out of five from 1965 to 1966 and two out of five from 1966 to 1970 with exception of a part of 1966 when only one out of four were scientists. Also except for a brief period of 1977-1978 there were no independent scientist member.[232] This trend still continues. The commission, by norm has former chairmen as members. A large Indian scientific dysphoria still has no say in the workings of the department of atomic energy. The actions of the departments are cloaked under strict secrecy clause. This excludes any meaningful analysis of its achievements.

Tatas have an excellent history of their philanthropic work in support of science, in addition of their contribution to the industrialisation of India during the Imperial rule. However, these are not above self-interest as can be seen by many examples. The Tata trust did support Saha in his scientific work before Bhabha appeared on the Indian scene. However, after Bhabha's institute was established, further grants to Saha ceased.

One must not separate the philanthropic work from the commercial interest of the Tatas. With minimum investment, Tatas got their name attached to a premier scientific institution of the government for eternity. Tatas funded the setting up of National Chemical Laboratory, Pune and wanted it to be named after them but it was not agreed to.[233]

In this connection one is tempted to recall the strenuous efforts Tatas made in getting commercial benefits as a precondition for their philanthropic proposals during the Imperial rule. Tata wanted to start an institute for scientific research and education. Deepak Kumar has described in detail how J. N. Tata sought to get exemption from the General Statute Law against the perpetuities for the benefit of Tata's descendants in leu of funding a research institute. The government did not agree but Tata tried to get the concessions from the secretary of state in London. Lord Curzon did not agree in spite of pressure from the press and offered 2000 pound yearly as assistance to the institute. Tata, however refused. Curzon left India in 1905 and J. N. Tata passed away without seeing the institute come to light. The institute, Indian Institute of Science, Bangalore, came into being in 1909 with assistance from several sources. The institute, despite Tata's efforts, could not carry his name. It is clear that trust's philanthropy is not always without self-interest.[234] *{Note-15}*

Thus, the proposal of a new institute for fundamental research was more for rehabilitating Bhabha. It was not an act of purely supporting Indian science but was also of very personal gain. It is surprising that Pt. Nehru could not see through the game. Instead he was sufficiently enamoured by the charm of Bhabha to give him unparalleled power and freedom to usurp the scientific agenda of the country.

7.2 Cosmic Rays to Atomic Energy

Let us now examine the goals set up by the founding father of the atomic research in India. Was he clear in his approach? Bhabha highlighted the cosmic ray study at the formal inauguration of the Institute on December 19, 1945, stating that [235]

"In cosmic rays nature has provided us with the biggest atom smashing instrument in the world, and whole surface of the earth is our laboratory"

On January 1, 1954, at the foundation stone laying ceremony of TIFR, Bhabha again expressed his opinion that the cosmic radiation provides us with more energetic particles than any other accelerator which can be used to study the same phenomena at significantly less cost. Bhabha again came back to justify the use of cosmic rays, saying that:[236]

"Now the accelerators are costly things. .... This would normally put them outside the scope of what we in India would do with limited finance. However, thanks to nature we have cosmic radiation that provides us with particles that are even more energetic than can be provided with any other accelerator. These can be used to study the same phenomena".

In enhancing the relevance of research in cosmic rays to nuclear physics, Bhabha was probably downplaying the importance of particle accelerators citing cost involved and paucity of funds. Bhabha's proposals for the new institute did not consider particle accelerator to be necessary.[237] Jahanvi Phalkey has discussed Bhabha's shifting interests beginning with cosmic ray physics, through research reactors, and finally fusion research for the period between 1952 and 1959 and its effect on the institute.[238]

His attempts at justifying cosmic rays for nuclear studies, notwithstanding why Bhabha actively and resolutely pursued a programme of accelerator building at the TIFR, is a question that Jahnavi Phalkey tries to answer.

"The other argument Bhabha made, was concerned more with the pursuit of his own research agenda. "The study of cosmic radiation forms the main field

of experimental research at this institute, though I hope that in the near future experimental work will also extend to nuclear physics. The two branches are very closely knit and indeed the elucidation of an important problem in nuclear physics, namely the origin of nuclear forces, owes its existence to the discovery of the meson in cosmic radiation".[239]

Bhabha found a linking hook where his own interest in particles and cosmic ray research, the agenda of carrying out fundamental research in the new institute and the referent of nuclear physics converged: in the search for the meson".[240]

Bhabha seriously began thinking, later, of particle accelerator,[241]

"The experimental research at the moment is mainly in cosmic rays, but I have no doubt that in the near future it will expand into nuclear physics, and in time the Institute may get large equipment such as a betatron, cyclotron and/or a van der Graaff generator... I would like to know if in your opinion the proximity of the sea is likely to make the site unsuitable for the operation of instruments like a cyclotron". Bhabha had begun to seriously consider the use of particle accelerators for experimental nuclear physics like other laboratories of the time".

The Atomic Energy Research Committee (AERC) meeting on May 15, 1946, recommended cosmic ray research at Calcutta, Aligarh and Bombay. A betatron was sanctioned to be established at TIFR, however this was not perused. In a meeting of the AEC meeting on Aug 15, 1948 to a query by Pt. Nehru, Blackett's advice was cautious;[242] *{Note -16}*

Bhabha began scouting for a linear accelerator and Raja Ramanna was put directly in touch with person in charge at Harwell and order was placed with Philips Electrical (India) Ltd. Ramanna was leading the nuclear physics group housed within TIFR.[243]

"Cosmic ray physics, nuclear physics and construction of electronic equipment were counted as experimental work. However, cosmic ray research dominated experimental work up until the arrival of the cascade generator in 1953".[244]

TIFR was now engaged in nuclear research for AEC. Nuclear physicists were divided between the priorities of nuclear research at TIFR and the experiments planned for reactor work. Conflict between the fundamental research and applied research began to take shape over the use of the Phillips machine. [245] {Note -17} To resolve the conflicts Bhabha requested Mark Oliphant[246] an Australian nuclear scientist, closely associated with Manhattan project, to evaluate the organisation of work in experimental physics. Oliphant was very critical of the scientific practices.[247]

> *"A month later, he submitted a scathing critique of scientific practice at TIFR. First, he noted the tension that arose from the TIFR hosting AECI's groups – resonating Kondaiah's complaint. An experimental physics group, very much like a group in theoretical physics, Oliphant argued, had to be organised around individuals. The laboratories of the AECI on the other hand would have to be organised around specific tasks and these two could not be conflated. Oliphant reminded Bhabha of issues not far from those Bhabha had raised with Saha only a few years ago, "The successful pursuit of an experimental program involves far more than definition of the experiments to be performed and the equipment to be employed. Such work must be planned in detail, using the fullest cooperation with theoreticians and with masters of various techniques in all groups. Leadership requires certain ruthlessness in discarding unproductive lines of experiment, as well as qualities of discrimination in the choice of experiments as skill in the employment of techniques. I am not at all convinced that these three qualities are present in many of the leaders of your existing divisions". Oliphant asked Bhabha to keep the groups small and try recruiting or training men of "the right calibre" before embarking on plans that are more ambitious".*

The very leaders about whom Oliphant gave such acerbic opinion went on to lead the atomic energy programme.

Ambiguity of TIFR within the larger agenda of AEC was resolved by a separate research facility which became possible due to Anglo-American approach to sharing nuclear technology that made it possible to set up a research reactor.[248] Atom for Peace, a United States initiative opened the gates of Nuclear co-operation. In 1955 AEET began constructing 1 MW research reactor with British Assistance. The reactor became critical two years later and was named Apsara, a consort in the abode of Lord Indra. Indian science could not escape a touch of the divine.

The establishment of AEET, Trombay, with one reactor and a training school (started in 1957) in place, raised a question on the necessity of the programme at TIFR. [249]

"If the groups' main purpose was one of training personnel, the AECI now had its own program – if the purpose was research, the AECI always had its own machines. This ambiguity around the groups' function was an expression of the ambiguity of the relationship between the TIFR and the AECI".

The trajectories of TIFR and AEET followed different directions even though a symbiotic relationship did continue for some time with the ties weakening with passage of time. The transmutation was beginning to take shape.

Bhabha needed trained manpower for his institutes. Bhabha's views on manpower were,[250]

"What we require in India today are people who are really on top of a certain branch of a subject, however narrow, rather than people who have a smattering of a large number of subjects but are incapable of doing any first-class work in any of them on their own".

Bhabha went about scouting for specialist internationally in various narrow fields and offered them positions. However, these few specialists could hardly deliver the promises without a dedicated workforce. With the quality of students coming out of the universities continually declining he probably had no option but to have inhouse training. In 1957 he started the Atomic Energy Training School (AETS). The training school provided him the required labour force to fulfil the later part of his above-mentioned statement, namely *"incapable of doing any first-class work in any one of them on their own."* These trainees became the backbone of the expanding conglomerate.

Bhabha probably aimed in the training school to produce scientist-equivalent of the central civil services who would have working knowledge of various fields so that they can be fitted in any area of research. These generalists provided the work force for the department's programmes. The teaching at the school was only marginally better than that offered at the universities. There was no intent and a scope for teaching the future scientists of the basic tenets of science leave apart the history of science which the students had never before glimpsed during their entire graduation years. The trainees, after completion of the course are placed directly on to the job where they pick up the mannerism and the ethos floating around. The strictly hierarchical set up forbids any dissent or frank discussion even on scientific matters. A fear psychosis guides the behaviour. It is no surprise that the manpower employment and training policy has failed to produce even a single world-renowned scientist. Even those specialists, with narrow field of specialisation, who were inducted in the department were not able to make any impact on the world scientific scene. Scientists in high administrative positions in general are more known for the position that they hold within the country and are unknown outside, in the world of science. One might find few exceptions only to prove a rule.

Bhabha was a perfectionist. His imprint was visible in everything that he did, from planning the layout of the institutes, design of buildings, lay out of gardens etc. He left very little for others to do. He was known to become furious if a thing is not done as per his thinking. This prevented others from applying their mind to the problem. This work-culture percolated down and many lesser capable leaders, in trying to emulate Bhabha, further eroded the eco-system.

7.3 Genesis of Institutional Ethos

Ethos, as per the Cambridge dictionary, are a set of beliefs, attitudes, habits, etc. that are characteristic of a person or group that gives it a value, a character. Ethics, derived from ethos, comes into being from individuals who make difficult choices, good or bad, right or wrong. It is these choices that chart the genesis of institutional ethos.

Pre-independence research institutes were built with private munificence around the university professors resulting in a close-knit community of research students around the leader. The ethos of research was instilled by the leader who lovingly nurtured his wards and passed on to them the values ingrained by him through his own growth under his teachers. Gradually the students grow and emerge from the shadows of their leader and the torch is passed on to the next generation of students joining the institute. A culture begins to be built around in the institute and norms are unconsciously and subtly established. Most such pre-independence institutes grew out of the university ethos by singular efforts of a Saha, a Sahani or a Bose and many others. In this scenario in Abdus Salam's advocacy of building institutes around "tribal leaders', the government found resonance with Bhabha emerging as the tribal leader for atomic energy. Bhabha, being brought up in a west oriented industrialist family with western education unconsciously, adopted a sleek corporate culture for the governance and functioning of the institute. The environment for the new institute could not be that of a university.

Bhabha was largely educated in England after his initial schooling at Bombay. He graduated and engaged in research in one of the best universities in Europe. He was seeped deeply in the western art, culture and the ethos and the scientific tradition. As a scientist he could take on his seniors like Blackett and engage with him in frank and fearless discussion. It was expected that he would bring along the same values, so central to the western science, that guided his behavior while in England, to India. Alas, Bhabha in India was a different scientist.

The manner in which Bhabha emerged on the Indian nuclear scene and eclipsed all others could not have been lost to the emerging scientific community of his time. Bhabha was an aristocrat, connoisseur of arts, music and culture and a diligent scientist, though his personal contribution to science unfortunately ceased early. His stature, position and power, vis-a-vis other tribal leaders, was incomparable. Bhabha got the freedom and the means to setup an institute of his liking. The manner in which he went about achieving his objectives must have left an impression on his colleagues about the means to be adopted in meeting their objectives. It is but natural that he is taken as a role model by the aspiring research workers of the time. One would have expected that Bhabha's interaction with the world leaders of science like Walter Heitler, Wolfgang Pauli, Powel, Enrico Fermi, Blackett and others at Cambridge would have given him a sense of what constitutes a free and fair exchange of ideas between scientists. His famous repertoire with Blackett, eleven years his senior and an established scientist, on meson would have certainly given him insight into respect towards juniors and the value of free discussions. He would have appreciated enough the value of disagreements in scientific arguments to instill the same values in the institutes that he was so lovingly and deftly crafting. In this setting it is pertinent to discuss Bhabha's actions in dealing disagreements with some equally talented colleagues, the result of which was their unceremonious exit.

Bhabha's dealings with some of his eminent scientific colleagues, Prof. Piara Singh Gill,[251] Prof. D. D. Kosambi[252], Prof. K. S. Chandrasekharan[253] and Prof. E. C. G. Sudarshan[254] is well documented. These renowned contemporaries who worked at TIFR are amongst some of the scientists who left the institute or were sacked. A case of a junior scientist K. A. George, who was punished for expressing his opinion about his deputation abroad and the action that followed, is forerunner of things to become norm in the Department of Atomic Energy (DAE). The manner in which these eminent scientists were treated points to the direction in which the value system that was developing.

Piara Singh Gill

Bhabha was a theoretical cosmic ray physicist with no experience in experimental work. He did start experiments using balloons while at Bangalore. Two years younger to Bhabha was another cosmic ray physicist, Prof. P. S. Gill. He was a self-made man who rose from a humble background to be recognised for his contributions to cosmic ray physics.[255] Gill was an experimentalist and had established a name for himself in studying azimuthal distribution of cosmic rays, spin of the meson and burst emission of cosmic rays. Gill returned to India and started working at Forman college, Lahore. His biography throws light on some aspects of Bhabha's relations with his contemporary scientists that has bearing on the ethos of the institute he crafted. Gill, admiring Bhabha for his contributions, writes[256]

> "....... *During my student days at Chicago, I had admired Bhabha for his theoretical contributions on the production of cosmic ray showers. I met him for the first time in Bangalore in 1940, when he was a Reader in Physics at the Indian Institute of Science Dr. Bhabha asked me to give a talk on bursts, the well-recognized work that I had done with Dr. Schein. After my lecture, Bhabha suggested that I go back to the States, as there would be hardly any future for me in India.*
> *I took this as friendly advice; however, the events which followed made me suspect that Bhabha was keen to keep me out of India for other reasons."*

Notwithstanding this advice to Gill in 1940, Bhabha did try recruiting him for TIFR on more than one occasion but he declined the offer.[257] Gill was studying distribution of cosmic rays at high altitude for which he was using Royal Air Force

(RAF) plane. Bhabha also was interested in high altitude research and he wrote to Air commodore Wheeler;[258]

> *"… …which stated that his Institute was the only one doing worthwhile cosmic ray work, and he would like to have a plane at his disposal for high altitude studies."*

It is clear that Bhabha wanted only TIFR to be recognised as the sole institute for cosmic ray research. Bhabha's request was declined, and he was asked to co-ordinate with Gill.

Bhabha was not averse to seek cooperation from Gill to represent India in commonwealth conference[259]

> *"In April 1946, a year after my expedition to study the production of mesons as a function of altitude, I received a cable from Bhabha. He was to represent India at the Commonwealth Scientific Conference, and I was asked to send my unpublished data for him to present at the conference on behalf of India. With pleasure I complied".*

However, he failed to acknowledge Gill's results.[260]

> *"……… I shared my results with Dr. Bhabha. Meanwhile, his Institute made some measurements with the help of an American plane at Bangalore. In acknowledging receipt of my results, he wrote to me on May 3, 1946, "…. We recently detected the same hump at 500 millibars in the curve of mesons penetrating 30 centimetres of lead. You will see this in the paper which will appear this month in the proceedings of the National Institute of Science." In that paper, however, no acknowledgement was made of my results".*

This sort of unethical behaviour did not stop here. Gill further writes; [261]

> *"In 1946, the Government of India selected six scientists including myself to visit laboratories in Europe and America for a period of six months. I decided to go to the United States to extend my experiments at various latitudes and heights, an idea I had been entertaining ever since the Tata Institute made a request to the Royal Air Force… …".*

"Prof. Compton, who was now Chancellor of Washington University in Saint Louis, called to invite me to give a lecture to their Physics department on my work in India. In Saint Louis, I stayed in the Chancellor's beautiful home and he attended my lecture. I spoke of my work in the Himalayas, particularly the measurements I had made in the airplane. I referenced Bhabha's work rather extensively".

"The evening after dinner, Prof. Compton told me that Bhabha had been in Saint Louis ten days earlier and had spoken on the same subject without mentioning my work at all. Bhabha also tried to persuade Compton to advise me to join him in Bangalore. After listening to talks by Bhabha and me on the same subject, Compton was of the opinion that India was a large country, and I should stay as far away from Bhabha as possible".

Gill's fate willed it otherwise. The country was going through the pangs of partition and situation at Lahore was increasingly becoming riotous. Gill was forced to relocate from Forman college, Lahore.[262] Gill met Pandit Nehru and was appointed as professor of experimental physics at TIFR.[263] Gill started balloon experiments from Juhu airport, and his work was noticed by the local press. [264]

"In Bombay I set to work with new zeal studying cosmic rays using balloons procured with funds from a grant from the Education Ministry of the Government of India. With the assistance of a couple of students, we started sending up balloons from Juhu airport.

"The work attracted the attention of the press, and a few write-ups appeared in the Bombay papers. Dr. Bhabha had meanwhile gone to England, and Mr. Choksi was the interim Director in Bhabha's absence. Mr. Choksi wrote to me that "no publicity could be given except through the authorities of the institute." I wrote to Bhabha about the unhappy situation. He wrote and assured me that I should not worry about it. ...."

Mr. Choksi was a professor of English at a local college and a close confidante of Bhabha. The irony of the situation should not be lost that a supposedly premier institute of fundamental research where a non-scientific person sat on the council, acted on behalf of the director and offered judgement on the scientists. This continued even after Bhabha's demise.[265] This was not an isolated incidence of humiliation.

Gill was so frustrated that he ultimately decided to resign and wrote to Pt. Nehru. When no response was received Gill learnt that Bhabha had received the telegram from Bhatnagar to make arrangements for Gill's visit. Bhabha did not inform Gill. Gill went to Delhi and described his difficulties to the prime minister. The prime minister called Bhatnagar and asked him to accompany Gill to Bombay, call for the council meeting to accept the resignation. The meeting was held in Tata House with Saklatvala in the chair though Bhabha was present. Saklatvala, accepting the resignation, asked Gill to tender an apology to the Director of the institute. Gill refused and walked out of the meeting. The institute sent the registrar with a typed letter of apology to be signed by Gill. Gill refused to sign it and wrote his own apology. That same evening Gill took the frontier mail for Delhi. Next day he briefed Pt. Nehru of the proceedings.[266] *{Note -18}* Gill was appointed as Officer on Special Duties (OSD) under the Atomic Energy Commission with a promise of increased emoluments, but the appointment letter indicated no increase. Bhabha's ire was not yet satiated. Later Krishnan (K. S.) told him that Bhabha opposed any increase.[267]

One can immediately infer that Saklatvala, presiding instead of the Bhabha who himself was present, seem to be a premediated act for humiliating Gill. It should be apparent to everyone that almost all prominent personalities who helped Bhabha in his mission of creating a niche for himself were made members of the council. The reasons for Pt. Nehru to ignore such transgressions are difficult to understand.

As OSD Gill did not have a laboratory, equipment or workers which added to his frustration and he left DAE to join Aligarh Muslim University (AMU) in 1948. Bhabha's harassment of Gill did not stop here. J & K government and AMU decided to establish at Gulmarg an observatory for studying cosmic rays with Gill as its Director. When request for help in establishing the observatory was sought Dr. Bhatnagar on behalf of the Atomic Energy Commission stated that the location for the observatory is not suitable for high altitude research and that the prime minister has declined financial assistance on the advice of the experts. [268]

It is interesting to see the forces playing at the highest level of science administration that was in the hands of Bhatnagar and Bhabha. The balanced response of the then Vice Chancellor of AMU, Dr. Zakir Hussain, to the denial of assistance for setting up of the observatory by the government leaves it in poor light.[269] *{Note -19}*

Gill survived and continued his scientific work and remained director of the observatory from 1951 to 1971.

Later Bhabha also wanted an observatory for high altitude cosmic ray research.[270]

".... that although he wanted to set up a station in Gulmarg, he did not want to cooperate with the already existing laboratory. It was surprising that he would consider Gulmarg at all, and I wonder how the Atomic Energy Commission could reconcile this with the views held by the government experts.

In June of 1954, the Prime Minister wrote to me: ".... I am greatly interested in the work that you are doing.... There are some proposals also for a high-altitude cosmic ray station to be put up on behalf of the Atomic Energy Commission. I do not think it has been decided yet where to have this. In any event, the work that you are doing must be continued and helped in every way".

In 1963 Bhabha, started High Altitude Research Laboratory (HARL) at Gulmarg, the same place that he had declared unsuitable when the question of setting up Gill's observatory was being discussed in academic and government circle. It is therefore clear that Bhabha was not as much interested in cosmic ray physics as he was interested in consolidating his position in science administration. He did not hesitate to go out of the way to put impediments in the path of other scientists who happened to cross his path. Being a theoretician, Bhabha could have cooperated with the experimental groups at Calcutta, AMU and Gulmarg. Which theoretical physicist would not like to have collaboration with strong experimental group in his field? Bhabha chose to ignore or subdue them. He later started his own experimental cosmic ray group with M G K Menon[271] and Bibha Chaudhury[272] working in the Kolar Gold mines. At a time when scientific research worldwide was transforming itself from individualistic to co-operative research such an attitude set the tone for nascent modern scientific research in India.

It is seen from the events related to the administration of TIFR that the scientists at the premier research institution were made subservient and answerable to the dominant non-scientific members from the Tata trust on the institute council. Raju writes about the new relationship after independence between industry and the state funded institutes;[273]

"It is routine to pass on the results of state-funded research to private capital, but Nehru built a unique system of science management in India, by going a step further and giving representatives of private capital managerial control over state-funded research. Therefore, he forced scientists to be subordinate to capital".

D. D. Kosambi

Bhabha's tiff with Prof. Gill was not an isolated event. The other prominent scientist who was the object of Bhabha's wrath was polymath and polyglot Prof. D. D. Kosambi. Kosambi was the foundation fellow of the Indian Institute of Science, Bangalore.[274] Kosambi then was working at the Ferguson college, Pune. His papers on numismatics were published in Current Science.[275] Bhabha must have come in to his contact after joining the institute in 1940 as Kosambi was a frequent visitor to Bangalore. Bhabha invited Kosambi to join TIFR as professor of mathematics. Bhabha's proposals stated that he will be an asset to the institute with his knowledge of mathematics and would be helpful with statistics at the cancer hospital.[276]

Kosambi became the founding member of TIFR. In the initial years Kosambi was quite close to Bhabha and often officiated as director in his absence. Relations between them were professional and cordial. Bhabha thanked Kosambi in his papers for mathematical and statistical help.[277] Kosambi, even after joining the institute, continued to work in diverse fields for which he was being recognised. Bhabha on the other hand was increasingly getting involved with administration. Kosambi disapproved this as he thought that a first-rate scientist should not get involved in institution building and stop research.[278] The relations between them started going downhill over time. The mathematics department at the institute had a batch of fresh scientists amongst whom was Dr. K. S. Chandrasekharan. Chandrasekharan became a close confidante of Bhabha and eventually rose to become the Deputy Director (mathematics). Kosambi was now side-tracked.[279]

Another factor that could have contributed to Kosambi's estrangement was the review, that he wrote, of Nehru's book the 'Discovery of India', pointing some defects in historical and political viewpoints.[280] Kosambi's Marxist views were known to all but his participation in the in the World Peace Movement against the atomic bomb, though initially supported by the Indian government, changed subsequently as the communists adopted the policy of confrontation against the Nehru Government.[281] Kosambi knew that his participation in World peace movement may cost him his position at TIFR.[282]

Bhabha was committed to atomic energy but Kosambi denounced the policy. Bhabha might have seen this as afront to him.[283] Way back in 1957 he wrote an article 'Sun or Atom' in which he discussed the merits and demerits of the two and concluded that;[284]

> *"But the huge primary source of energy today remains the sun. Direct utilisations hindered only by the desire for prestige, which makes India waste so much of her money in supposed research along other lines ".*

Kosambi again emphasised the need to work on solar energy. He could foresee the difficulties but was optimistic. [285] It is clear that Kosambi's strong advocacy of solar energy for country's energy needs might have put him at odds with Bhabha, a strong votary of atomic energy.

The difficulties for Kosambi were mounting. Kosambi refused to quit and Bhabha waited for an opportunity to come up.[286] The opportunity did come in the form of a paper Kosambi sent for publication on Riemann Conjecture, a difficult problem and an enigma in mathematics.[287] Kosambi chose to publish the article in Journal of the Indian Society for Agricultural Statistical (JISAS), a journal that did not peer review the submitted articles. K Chandrasekharan brought this to the notice of Bhabha and complained that,[288]

> '*Kosambi's half-cooked and imperfect research work caused damage to the reputation of his department.*'

Bhabha, instead of advising Chandrasekharan to scientifically counter the paper, did the rest. Kosambi was told not to re-join the institute after vacation. The response of the TIFR council was muted, [289]

> *"Kosambi was relieved from the institute. None of the members of the council except one spoke in favour of Kosambi..........*
> *Kosambi who was next only to Bhabha was not given even a simple send off when he left."*

There is no record of the reaction of the scientific diaspora of the institute. The institute's new building was ready but Kosambi was not destined to enter his designated room. He had probably outlived his utility to Bhabha.

A review of Kosambi's controversial paper, published in American Mathematical Journal, acknowledged that the article was of very high standard and it contained some novel ideas.[290]

"…in a review of this paper published in an American Mathematical journal the critic remarks, 'in his proof Kosambi has conveniently taken many things for granted.' Although Kosambi had failed to prove Ryaman's conjecture, his article contained many important novel ideas; but as professor Masani had said, unfortunately no one tried to follow and work further on them. Kosambi's attempt in connection with Ryaman's conjecture failed but the article itself undoubtedly was of very high standard".

Kosambi continued to work on Riemann hypothesis even after leaving TIFR and published another paper probably to clear misgivings and explained;[291]

"The method of proof for deductions based upon probability differ radically from those of pure mathematics. Conclusions cannot be 'true or false' without qualification".

Prof. Dani has aptly reviewed the work of Kosambi which shows his work on Riemann hypothesis in proper perspective;[292]

"Kosambi's last paper on the theme appeared in 1964, in which a proof of the Riemann hypothesis is again claimed, and this paper was also published in JISAS! It would be worthwhile to quote here in some detail from the review of this paper in Mathematical Reviews, that appeared after Kosambi's death, written by the well-known number theorist A Re'nyi: "The late author tried in the last 10 years of his life to prove the Riemann hypothesis by probabilistic methods. Though he did not succeed in this, he has formulated the following highly interesting conjecture on prime numbers." The reviewer goes on to describe the conjecture, comments that it would be even more difficult than the Riemann hypothesis to prove, and concludes with "Nevertheless, the conjecture is worthy of study in its own right, and the reviewer proposes to call it the Kosambi hypothesis" in commemoration of the enthusiastic efforts of the late author".

The ouster of Kosambi from TIFR was certainly not on his scientific capabilities. Another likely extraneous factor is indicated by Raju,[293]

"According to a well-known anecdote, at a social gathering in TIFR, speaker after speaker got up to praise the Tatas for their benevolence in contributing their hard-earned money for the noble causes of research and education in setting up TIFR. Kosambi, a Marxist, knew the real reasons why TIFR was set up. However, he disagreed only with the description of the Tata money as "hard-earned." He felt compelled to explain that capital often starts off as rogue capital before turning respectable. And he illustrated this by pointing out that the Tatas had earned big during the opium wars. The truth, however, may not be spoken loudly in a hierarchical system of management—Kosambi's remarks naturally did not go down well with Bhabha who was a close relative of the reigning Tata."

Kosambi had only stated a fact openly, as we know that opium was a major trade commodity during the reign of East India Company and Tata's along with many others did trade in it.[294]

After Kosambi's inglorious exit from TIFR, Director General of CSIR wanted to avail of his services. It was a tricky situation since Pt. Nehru was taking keen interest, after Bhatnagar's death, in CSIR and he would have certainly consulted Bhabha. The control on scientific institutions by individuals was all-encompassing,[295]

"Some satraps had established themselves in the field of science in India-Bhabha and his Atomic Energy Department along with all other related institutions on the one hand and on the other the CSIR and its network of National Laboratories all over the country."

Kosambi also found it difficult to obtain an institutional association, a precondition for the offer.[296] The path was not an easy one. The appointment could be made only in 1964 during Prime Minister Shastri's tenure.[297]

In the light of the rise of individual scientists to position of power, with state funds at their disposal Raju's comments though stated in relation to Kosambi's plight are largely relevant for post-independence science and scientists,[298]

"Post-independence, however, science became professionalized—a quest for funds, not truth. Post-independence scientists in India (think of any famous name) are people who control (or have controlled) vast funds and power. It is impossible to say how their scientific work, if any, has benefited the people in their everyday lives. There is a vast difference between J C Bose, inventor of the radio, and Raja Ramanna, Kosambi's junior colleague, and former chairman of the Atomic Energy Commission. On the other hand, scientific illiteracy ensures that scientific truth is decided purely by authority, making scientists even less accountable than bureaucrats, or businessmen, or the judiciary. Big funding with no accountability has led to the emergence of a veritable science mafia in the country, and Kosambi was one of the isolated few to resist it."

This also brings us to a question of resolving scientific disputes as there does not exist an appellate authority in science. Any point of contention in a scientific work is best resolved by scientists themselves through examining the findings and there should have been no occasion to pass judgement on extraneous grounds. The question that brings to mind regarding this episode is a very fundamental tenet of science. Why, with a talented pool of mathematicians in the institute the scientists who complained to Bhabha about the quality of Kosambi's work did not put their minds together to come up with a refutation of what Kosambi had published? This would have been the right course. Moreover, why Bhabha did not apply the western scientific ethos that he must have been imbued with, during his studies abroad, to tell his colleagues to refute it scientifically. Instead he used his administrative position and took recourse to oft-invoked idiom, 'sullying of the image' of the institution to relieve Kosambi of his position. Kosambi was denied even an opportunity to explain his work in any forum of the institute. When prestige of the institution becomes a sacrosanct entity the absence of scientific response mechanism often leads to high handedness and inappropriate action.

K. Chandrasekharan

K. Chandrasekharan [299] was working at the Institute for Advanced Study (IAS) at Princeton when Bhabha invited him to join TIFR in 1948. Bhabha gave him freedom to organize the mathematics activity. In a short time, the school of mathematics gained international reputation. He held important positions within

TIFR and was member of Scientific Advisory Committee to the Cabinet, Government of India during 1961-66. In 1965 he decided to leave the institute at the time when his efforts were bearing fruits. Seshadri writing an obituary for KC reminiscences; [300]

> "...... *In 1965, there was a rumour in the Tata Institute that KC was planning to leave the Institute. Very soon, it became a reality, and this came as a sudden and unpleasant surprise for many of us mathematicians at TIFR. The general impression that went around was that KC's leaving was because of his differences of opinion with Homi Bhabha on many issues, which perhaps was correct. In this connection, I should say that due to chance, when the furniture of the Institute was changed in some of our rooms, my old table was replaced by a bigger table. When I was transferring some of my old correspondence and other papers to the empty drawers of this table, I found two small pieces of paper with some notes in ink. They turned out to be parts of some exchanges between KC and Bhabha. Obviously, they were left there when KC's table was being cleared. One of them was a note signed by KC addressed to Bhabha and was about the closing down of the Institute canteen on a certain afternoon to accommodate a meeting of the Department of Atomic Energy (DAE). KC had written that some of the faculty members had brought this intended closure to his notice as a complaint and Bhabha had replied saying that he (KC) should bring to the notice of the complaining faculty that DAE is the financial supporter of the Institute and therefore there is no point in such a complaint being made. KC seemed to have started a retort but had given it up in the middle".*
>
> *"Perhaps there were very significant differences of opinion between them and it is possible they did not see eye to eye on some issues. But, one possible reason which I have heard was that KC used to spend the summer every year in Zurich, Switzerland, and he was a very good friend of Eckmann and others at ETH who wanted him to settle in Zurich".*

The cavalier response of Bhabha to the complaint from a colleague who happens to be holding the office of the Assistant Director and was an eminent scientist himself would have piqued even a lesser known scientist in that position. Bhabha himself was controlling the purse strings of DAE and was also the receiver of funds on behalf of TIFR hence this disdained response is not justified by any standard unless it is aimed for humiliation. The discord between the two could probably be explained by general malaise prevalent in post-independence institutes, aptly put by Raju in relation to Kosambi;[301]

"A hierarchical knowledge-management system does not allow the worker to be more knowledgeable than the manager, for this risk exposing the ignorance of the manager. To put Kosambi in place, Bhabha appointed Chandrasekharan, a junior and undistinguished mathematician, over his head as deputy director. It always seems a smart move for the manager to appoint a subservient deputy. To ensure constant dependence, sycophants are selected for incompetence. This can gravely damage the entire system through a chain reaction, leading to long lineages of sycophants of progressively degraded quality, institutionalising incompetence".

Chandrasekharan after joining TIFR, went on to build a strong group in mathematics and the institute gained international recognition. Bhabha on the other hand had ceased to contribute scientifically long before, as he was occupied with increasing burden of government work. Could this be the cause of friction between the two that led to KC leaving the institute? Was Chandrasekharan a victim of hierarchical science management where knowledge is also hierarchal?

E C G Sudarshan

We have so far seen the case of a cosmic ray experimentalist, Gill, who matched Bhabha's international cosmic ray standing but the two could not work together and complement each other's capabilities. Who was to blame? We have seen how Kosambi, an internationally known scholar was treated by Bhabha. The other case is that of Chandrasekharan, a promising young scientist brought in to probably put in place a capable senior colleague and who, after gaining recognition in his own right, became a person non grata.

We now focus our attention on how the research scholar E. C. G. Sudarshan and junior scientist K. A. George were treated. E. C. G. Sudarshan was a research student who joined the institute under Bhabha with a lot of expectation. However, Sudarshan soon got disillusioned as his guide was unable to give him time for meaningful interaction.

E. C. G. Sudarshan was a young scientist whom Bhabha invited to join him as he could not see research in quantum field anywhere else except his institute but could not retain at the institute.

" Sudarshan was born in a Christian family from central Travancore in 1931 (he would convert to Hinduism about two decades later). He received his master's degree from Madras Christian College in 1951. As he later put it, at the time, Homi Bhabha could not tolerate physicists studying quantum field theory anywhere in India outside of the Tata Institute of Fundamental Research, Mumbai. Given Sudarshan's interest in the subject and, more importantly, his promising mastery of it, Bhabha invited him to join them".[302]

Sudarshan worked in the cosmic ray group at TIFR as Bhabha's student, however he found that Bhabha does not have time for him. Sudarshan left TIFR after a couple of years work. The international fame that he subsequently achieved can be judged by two very illuminating papers by R. E. Marshak[303] and Luis J. Boya.[304] These papers give an idea of the loss suffered by Indian science. Marshak, in his paper, apologetically feels responsible for Sudarshan missing the Nobel prize.[305] Boya's biographical account of Sudarshan's life and work is fascinating story of a scientist who was nominated as many as six times for the Nobel prize.[306] Bhabha later tried to woo Sudarshan back to TIFR but it was probably too late.

This tragically is a widespread situation in which capable scientists, burdened with administrative work, are unable to provide meaningful guidance to research students leading to the frustrated student seeking scientifically greener pastures abroad.

Plight of a Scientist -Story of K. A. George

We now take up a typical case of a junior scientist in government service, that can showcase the eco-system that was beginning to take shape in a premier institute. It is the case of a TIFR scientist K. A. George who was sent on deputation in 1958 to Blackett's laboratory for training and research in fusion for a year. After about three months George wished to come earlier than planned as he felt that he has learnt whatever was there to learn and further stay will hamper the work back at the institute. His group leader, Phadke agreed and forwarded his letter to Bhabha. Bhabha did not comment on George's letter and instead questioned Phadke on the starting of fusion research and told him that no work can be started without the approval from the Director. A committee was appointed to discuss the progress of the accelerator work that was being done under Phadke' s charge. [307] *[Note-20]*

It is strange that Phadke was questioned on starting the work on fusion. As per the government procedure George's deputation and his subject of training would have been approved only at the highest level. Phadke would not have been in any position to send George on training abroad or decide on the subject for training without approval from the director. How would anybody understand Bhabha's action except that he was doing this only for slighting Phadke for supporting George. Even Blackett's suggestions about George's work was ignored. George stayed and was punished after his return. He was denied promotion; [308] The grounds for denying the promotion are nothing unusual;[309]

"The Common Room (faculty) had denied promotion for George on grounds that:

(1) he was not able to bring out any of the instruments to a stage where they could be used as tools of research.

(2) He has inadequate number of publications to his credit.

(3) His attitude towards his deputation abroad was unsatisfactory."

"George's was the only case where the promotion had been denied, which also meant no raise in his salary. He petitioned Bhabha. He first recounted his early work with group: "New ideas were presented during that period, ... in my Report (1) on the proposed 15 MeV Cyclotron.... It was declared policy in the Section that one's work would not be assessed on publications. Therefore, there was no encouragement for publications as such". [310]

This 'declared policy' is a problem that is unjustly faced by government scientists who have to develop the instrument either for their own research or for others work.

"In many cases, apparatus builders only hoped to become users of the apparatus they built. Building complicated apparatus for one, took several years, which meant that builders took time to publish observations of consequence to the community. Given the case of the builder groups at the

TIFR, they were but building prototypes for training purposes, familiarising themselves with the technique and technology of particle accelerators. That could hardly have qualified as original and publishable. This judgement, even of George alone, must have appeared rather unfair to the three teams".

George tried to explain the achievement in a new field and his efforts in preparing a report on plasma research. Phadke on the other hand reminded Bhabha about abandoning the cyclotron project and the fact that he was asked to find some more useful line of work. Their efforts to explain were of no consequence. To further rub the salt, M. G. K. Menon wrote to Phadke to give progress of work on Ven de Graaff and X-ray machines that the group promised to build. The consequence of this standoff was a joint response from the group. They expressed concern about their future in the group and asked for an enquiry into the work of the group. They highlighted the group's achievements which were appreciated by the Director on his visits. They explained how the 12'' cyclotron was dismantled after its completion so that the magnet could be used for plasma research. They also told about the policy that their work will not be judged on the number of publications. The rift between Phadke and the group came to fore as he failed to inform them of the faculty's opinions. [311] *{Note-21}*

Publication of papers or its absence is used very often, in research organisation, as a double-edged sword to deny a legitimate carrier progression to a researcher or an instrument developer.

George got punished for the afront of even expressing his opinion on his stay on deputation and Phadke got a rap for supporting his subordinate. The only positive aspect of this unfortunate incident was the combined representation by all the scientists of the group in support of a colleague and a faceoff with a powerful Director, a rare occurrence in government's institution.

To summarise, Gill, Kosambi and Chandrasekharan were well-known and established scientists. Gill was forced on Bhabha while Kosambi and Chandrasekharan were brought in to boost the image of the newly formed institute in the initial stages. Their attaining fame, through their diligence, was probably their undoing. The message that Bhabha probably wanted to give to other scientists was indeed delivered. Sudarshan's plight was the plight of a young intelligent man looking for guidance and intellectual interaction with his guide, the time for which unfortunately his guide could not find.

What does a scientist need in an eco-system apart from humility, honesty and transparency in decisions and ethical behaviour in scientific matters? He needs ability to communicate and interact on equal terms. Shiraz Minwalla, a scientist at TIFR, in an interview after receiving ICTP 2010 prize, has succinctly put it as follows; [312]

"What is not as good in India—and this is hard to fix—is the quality of interaction you get. At Harvard, the general level of discussion is higher than at TIFR. For instance, at Harvard I would go to lunch every day and we would always have an interesting conversation on the latest papers on the Internet. That somehow does not happen at TIFR, for various reasons: for instance, there are fewer postdocs of the same level. The second thing is that a place like Harvard, or Princeton, gets a continuous flow of extremely exciting people coming to visit. At any given week at Harvard there would be at least three visitors, each of whom has something completely new to tell you. India is certainly not at the centre of the traffic of physics in that respect. If you are at a place like Princeton or Harvard, you stay current just by being there; it takes no effort to keep track of everything that is happening in the field".

If this is the condition in the crown institution of the DAE one can imagine what ecosystem exists in its other institutes. It is a sad commentary on Bhabha's ambition of creating a Cambridge in India.

Bhabha was supreme not only in the affairs of the department of atomic energy but also in other areas of government scientific activity, being a member or chairman of innumerable committees and commissions. He managed the affairs of Atomic Energy like an autocrat. He knew that with the department covered by secrecy laws and with Nehru on his side he is answerable to none. The question that arise is why Nehru was so powerless as not to confront Bhabha on the problems faced by Prof. Gill, a person whom he appointed knowing fully well Gill's international standing and his contribution to experimental cosmic ray physics. Instead of confronting Bhabha on the problems faced by Gill he appointed him as OSD knowing fully well that Gill could not have lived without a laboratory for long. He sided with Bhabha without realizing the harm being done to scientific ethos. Had it been in any western university both would have found their place on the merit of their scientific contributions. In government scientific institutions, on the other hand, balance tilts in favour of the person holding higher administrative position to the disadvantage of hapless subordinate. The attitude of this individual has a large bearing on the eco-system.

Bhabha was an intelligent, well read and multi-faceted personality having interest and indulgences in arts and music besides engaging in science. Nothing much is known about his views about the freedom movement that has engaged a large number of fellow citizens of his time. This is not surprising because the house of Tatas was either leaning towards the then rulers or at best was neutral to the ongoing freedom movement. Nearest one can get to have a glimpse of Bhabha's political views or inclinations are from his letters as an intelligent well-read nineteen-year-old, to his friend Hormasji 'Homi' Manneckji Seervai,[313] who went on to serve as Accountant General of Bombay. Bhabha was a great admires of Napoleon. He expounds his views in a letter to him dated 2/1/1929. He agreed with Napoleon's views on liberty and the need to restrict it for some.[314] *{Note-22}* Towards the end of this rather long letter Bhabha writes, admiring Napoleon;

"I seriously ask you to read Napoleon by Emil Ludwig. It is great work of literature as an autobiography and will be highly instructing to you in political history. He was a great man. It will inspire you to work and labor hard. Give twofold strength and energy to you".

One can see how much Bhabha was influenced by Napoleon. Democracy is acceptable for want of an alternative and his views on liberty not being universal are astounding;

"'liberty is only the need of the few, the few whom nature has endowed with exceptional talents'"

and he agrees with Napoleon in restricting the liberty

"there is no danger in restricting it. The crowd loves equality.''

This largely explains Bhabha's autocratic behaviour and his attitude towards fellow scientists. This autocratic behaviour had a role to play in the eco-system in which the seed of fundamental research grew into a banyan tree of atomic energy establishment.

7.4 Marginalization of Universities

The most striking effect of the policy of the centralisation of government science was on the university education. It not only denuded the universities of talented teachers and researchers; their funds were also controlled by the government. This led to decline of higher education and research. Though centralisation was the government's policy, implemented zealously by Sir Bhatnagar, Bhabha could not be absolved of his responsibility. He did try to argue against drawing manpower form the universities but fell prey to temptation himself.

In 1946, the Atomic Energy Research Committee (AERC) was established under the aegis of CSIR.[315]

"The board met for the first time on May 15, 1946. Although little can be discerned from available documents about why Homi Jahangir Bhabha was appointed the chairman, ……… No less consequential for his choice was his connection to the Tata industrial family, given that J R D Tata was also a member of the Board for Research on Atomic Energy (BRAE). The meeting was held at Bombay House, headquarters of the Tata industrial establishments."

Saha was one of the members of the committee. The important decisions that were taken were,

"(i) the universities should be encouraged to impart elementary instructions in the theory and experimental techniques of atomic physics,

(ii) the existing centres of atomic research, namely the Palit Laboratory of the University College of Science, Calcutta, the Bose Research Institute, Calcutta, and TIFR, Bombay should be strengthened, and

(iii) TIFR should be made the main centre for all larger programmes of atomic research, until the stage is reached for a full-scale programme on atomic energy development".[316,317]

The committee also approved grants for Calcutta cyclotron, radioactivity research at Bose Institute and a betatron at the new institute, TIFR. There was no mention of Raman and R. S. Krishnan's plan at IISc, Bangalore, for nuclear research.[318] It is significant to note that a laboratory that was barely two year into its existence and was yet to establish its credentials was earmarked as the sole nuclear physics institute in the country.

"In the period between 1946 and 1948, Bhatnagar played an important role in ensuring that one laboratory emerged as the central laboratory of nuclear research, within the reigning CSIR logic of developing one good laboratory dedicated to one purpose – in fact the logic of the state, of concentration and nomination."[319]

Saha could manage to save his institute as a university laboratory but IISc, Bangalore failed to capitalise on the experience and skills of R. S. Krishnan for nuclear research.[320] Saha knew that his laboratory was spared as he began the construction of cyclotron under colonial rule and before the establishment of Atomic Research Committee.[321] Nevertheless, Saha proposed initiating work on Cockcroft Watson generator, linear accelerator and electron cyclotron and to train students. Bhabha responded by emphasising that all plants required for generation of atomic energy are the responsibility of the state. He approved the plan for electron cyclotron and suggested writing the proposals afresh. Saha's humiliation was complete.[322]

It is pertinent to note that Saha never required permission under the colonial rule to pursue his scientific activities. However, under the new dispensation, in free India, he was at the mercy of others who had much less experience in the field. With channel to the prime minister also blocked he appealed to the president of India who was visiting Russia, extoling him to see some of the laboratories and see how capable men of science are supported with sufficient funds. His efforts however went in vain.[323]

The case of IISc Bangalore's proposals, for starting research and teaching programme and its rejection by the Atomic research committee led by Bhabha is a clear example of how powerful individuals who control the funds pursued their own agenda to the detriment of university science. Raman wanted to start nuclear physics in his department. Krishnan (R. S.) had, by this time, five years of research experience and had obtained his DSc degree for his work on artificial radioactivity in 1941. Raman held high opinion about his experimental capabilities.[324] In his thesis he claimed discovery of nine isotopes, published twelve papers, all dealing with deuteron bombardment of heavy elements including uranium and thorium…….. Of these, he published three on his own, one with Norman Feather, two with Gant, three with Banks and four with Nahum. They published four of these in Nature, six in the Proceedings of the Cambridge Philosophical Society, and three in the Proceedings of the Royal Society.[325] Krishnan (R. S.) returned to the institute in 1941 ready to follow-up his research at Cavendish.[326]

> *"In March 1942, Krishnan submitted a proposal entitled "A scheme for power production from uranium fission". On his copy of the proposal, Krishnan noted, "Submitted in March 1942 and withdrawn after consultation with Sir C. V. Raman and the Director ……… From available sources, it is not clear if he discussed the proposal with the newly appointed professor, Homi Bhabha…………".* [327]

Writes Jahnavi Phalkey and leaves the questions: Did Bhabha know of the IISc proposal, in the realm of speculation.[328]

However, Jahnavi Phalkey rather obliquely confirms Bhabha's knowledge of Raman and Krishnan's project and writes, though in relation to Bhabha's lack of realisation of importance of particle accelerator for nuclear research;[329]

> *"It was four years since the Dorab Tata Trusts had made their first grants to Bhabha for cosmic ray physics work in Bangalore and an equal number of years since their grant for the Calcutta cyclotron. Bhabha was certainly aware of the project, and quite likely the proposal written by Raman and Krishnan while he was at the Indian Institute of Science, Bangalore. It is not that remarkable that a cosmic ray physicist did not consider particle accelerators as necessary apparatus for nuclear physics research at this juncture, but it is important for the purposes of this dissertation that Bhabha's plans for the institute did not carry a proposal for any such installation."*

> *"Along with Raman, and Krishnan, Bhabha must have understood the significance of the proposal."*[330]

In January 1947, R S Krishnan again submitted the proposals for nuclear research to the Atomic Research Committee whose members were Bhabha, Saha and Krishnan (K. S.). Raman emphasized Krishnan's (R. S.) expertise in the field and told the committee of having sufficient funds for the new activity. However,

Bhabha argued that the funds for the institute came from the government and the institutes own plans cannot be drawn in isolation of the wider policy of centralising the research at one institute.[331] *{Note-23}* Saha did not attend the meeting. Apparently no decision was arrived at. The type of conspiring that went on to deny nuclear physics research in a University is clearly established by Jahanvi Phalkey;[332] *{Note-24}* Bhatnagar felt that the institute should not be allowed to create a chair in nuclear physics as it will conflict with the development of Bhabha's laboratory. Bhabha requested Bhatnagar to convince Saha to vote negatively. Bhabha also suggested that the move be suppressed through the representatives in the institute council, arguing that it would be an unnecessary duplication and unwise use of philanthropy. Bhatnagar was obliged to Tatas for financial support for CSIR laboratories and thus he was keen to ensure a prominent position in nuclear research to Bhabha's laboratory.

In the meantime, Bhabha and Saha found their place on the IISc council as a nominee of the Tata family and of the Ministry of Education, Government of India respectively. Since both were also members of AERC there was little chance of Krishnan's (R S) proposals being accepted. 'Bhabha's decision embodied his position with the state (as chairman of the AERC), his relation to the patrons, the Tata family, and his position as a physicist. He had Bhatnagar's support and neither Saha nor K. S. Krishnan attempted to support Raman and R. S. Krishnan's proposal.[333] No decision however, was taken in the meeting.

Bhabha, with the help of H. J. Taylor, a physics teacher at Wilson college whom he had co-opted as experimental physicist at TIFR, prepared a report on Krishnan's proposals. Bhabha being the architect of the AERC policy on scientific research, the report was to arrive at the pre-decided conclusion of consolidating all nuclear research at one institute namely TIFR. It is interesting to note that the report needlessly emphasises the obvious, namely that the nuclear physics is not the same as acoustics and optics,[334] unless it was an attempt to rub salt and belittle Raman. It would be naïve to think otherwise.

A meeting was held at Wilson college, Bombay to consider the proposals and Bhabha-Taylor report. H. J. Bhabha, J. C. Ghose, Director IISc, Bangalore, H John Taylor, C. V. Raman and R. S. Krishnan were present at the meeting. Saha, J. C. Ghosh's classmate and Krishnan (K. S.), student and co- discoverer of Raman effect did not attend probably to avoid unpleasant faceoff with Raman. Saha could have argued forcefully for a mandate for nuclear physics research and education within the university system but that was not to be.[335]

In 1948 Atomic Energy Commission (AEC) was formed and Saha, who was member of AERC, was replaced by K. S. Krishnan. The commission under Bhabha dispensed with the first recommendation of AERC regarding involvement of universities.[336,337] The committee decided that all large-scale research in atomic physics be concentrated in one centre and that centre was identified to be TIFR.[338] Raman was not happy with the situation. He submitted a memorandum to the government of India clearly voicing his apprehension that the exercise is to create a monopoly in the subject and favour certain institutes to the exclusion of others. He feared that this will result in starving out everyone else from the field.[339] Raman's apprehensions were indeed prophetic.

Bhabha emerged as the sole administrator of nuclear research within the country. Soon he was also recognised as the sole representative from India when he was requested to nominate nuclear scientists for Atom for peace meeting in Geneva. Department of Atomic Energy (DAE) first asked for nominations from Institute of Nuclear Physics (INP), Calcutta, only to summarily reject the nomination, denying Saha's group any representation on a specious argument that the meeting was about nuclear technology and not nuclear physics, an unsustainable distinction.[340] It is pertinent to note that Bhabha's own institution would not have stood the same criteria at that point of time.

It is clear that the political leadership of that time, particularly Pt. Nehru, placed too much faith in an individual who worked for his own benefit and whose control on funds for nuclear research resulted in marginalisation and weakening of nuclear physics research in teaching institutions. Had there been a single unified society of scientists, as Rajagopalachari was trying to forge, advising the government on such matters, the political leadership and hence the scientific growth would not have fallen prey to couple of individual's own ambitions. Saha did raise critical issues with Pt. Nehru in the parliament and in open forums but was ultimately marginalised. Saha also could not rise above personal animosities at critical junctures and did not support Raman's proposals for nuclear research at IISc.

We are now at the end of the chapter and are confronted by a counterview[341] to Gandhi's quote with which we started this chapter.

"Most schools of thought accept a sharp dichotomy between ends and means; and discussions about means are always related with their moral

implication and property, or about the extent of their theoretical and contingent compatibility with desired ends. It has been observed that in the western tradition there is a tendency of claiming that the end entirely justifies the means – moral considerations cannot apply to the means except in relation to ends".

Do we get an answer? There is no record, at least in public domain, of Bhabha ever meeting Gandhi. His views on Mahatma Gandhi and his philosophy are not known. An admirer of Neapolitan is unlikely to appreciate Gandhi. Is it, that the west-trained scientists, particularly Bhabha, at the helm of post-independence scientific affairs followed the above dictum?

*

8

Other Debilitating Factors

In addition to the centralisation of research, concentration of power in the hands of few scientists and personal and over reliance of the then government on individuals, the ethos in science eco-system were also affected by some additional debilitating factors. These are enumerated below.

1. The relationship between the political class and the scientist.

2. Appointment of eminent scientists to head the new institutions that had their core objectives far removed from the expertise of the new appointees, often a case of square peg in a round hole.

3. The cavalier manner in which the first National awards were bestowed, on some favoured government scientists ignoring other deserving scientists.

4. The tendency of power wielding administrators to favour individuals who can be termed as the 'Blue eyed Boys' the practice of which spoils the working atmosphere.

A brief discussion on these is presented in this chapter.

8.1 The Scientists and the Political Bosses

Early indications of the nature of the opinion of Indian politicians towards Indian men of science and scientist's response to the politicians can be gauged from examples namely of Saha and Bhatnagar. Saha was involved in science planning even before independence. He was the member of the committee that was formed to plan for the scientific infrastructure. He is credited with the planning of river management and the Damoder valley project. His uncompromising views on Charkha and Gandhi did not get him support from the Indian National Congress for the election to the first Constituent Assembly in 1947. Saha was critical of the government's record on education, industrialization, health, river valley development, and planning. He contested the election to the parliament as Independent candidate in 1952 and defeated the Indian National Congress candidate. Saha's tenor in the parliament was uncompromising on the government's science and industrialization policies. He continued to raise his voice in the parliament and outside, through his writings, in his journal 'Science and Culture'. His incisive questioning in the parliament and the response from K. N. Katju, the then Home Minister and previously Governor of Bengal showcase the disdain that the politicians have for upright scientists. Anderson writes; [342]

"At the outset of Saha's political career in parliament, K. N. Katju, whom Saha had criticised in 1938, warned him to confine his attention to the laboratory, saying that he did not really belong in Parliament".

This typifies the relation between a scientist politician at one end and the career politician having peripheral knowledge of science at the other. The antipathy of the rulers of free India towards scientists was no different than that of lord Curzon who, on the possibility of scientist on the council, had said; [343]

"Council is an administrative body. The place for expert is to advice, not to administer, and his place is accordingly outside it, not on it".

The ignorance or lack of intricate scientific knowledge of the politicians in power is often exploited by scientists to further their careers.

"Saha was one of the few who stood up to Nehru, particularly in question to atomic energy".

writes Anderson [344]

"Nehru was not always clear or well prepared and showed at times an uncertain or confused grasp of technicalities for which he was responsible".

Saha did not mince his words when it came to express his opinion on science and policy matters. It could probably be the reason for his being cast away by the political class from the national planning, especially in the area of atomic energy on which he held contrary views. His view was to work towards improving the industrial infrastructure before taking up nuclear industry. Saha was not alone in opposing the model for development of atomic energy as propagated by Bhabha. Kosambi also did.

8.2 Square Peg in A Round hole

The government of free India, in eagerness to expand the scientific and technological footprint of India, started a large number of research institutions but was faced with the problem of finding capable scientists for administering these. Since a large number of capable researchers were serving in the universities it was imperative that they be called upon to share the burden. The government inducted many scientists into the new institutes without much thought about matching the job profile with the new incumbents. Sir S. S. Bhatnagar, a Chemistry professor came to occupy a purely of administrative post which was to oversee the establishment of the new institutes. D. S. Kothari a theoretical astro-physicist from Delhi university became Director General of DRDO, and K. S. Krishnan a collaborator of Raman on Raman scattering as Director of N P L, Delhi. What impact this mismatch had on effective implementation of the core objectives of the institutes is a matter of separate study. One can look at CSIR as an example.

CSIR was tasked to setup five national laboratories to speedup scientific and industrial research. National Physical laboratory, Delhi (NPL) being one, Krishnan was appointed as its first director. Krishnan, was persuaded, during the science congress, in 1947 by Nehru to take charge of NPL [345]

The core objective of NPL is described by Sir S Bhatnagar as; [346]

"...the laboratory's foremost function will be the maintenance of fundamental and derived standard, and the undertaking of research with a view to achieve greater and greater accuracy in the measurement of those standards,".

Krishnan had worked in the field of light scattering, magnetism and material science and had no work connected to research or maintaining the measurement standards. In addition, contrary to the core objectives, NPL was tasked to develop a solar cooker, an item that had nothing to do with the task assigned to the institute. The result was an utter failure. The scientists spent considerable time and energy on its development, and it was announced with much fanfare by national broadcaster. A unit was presented to the then prime minister. The cooker was commercially produced by a steel utensil manufacturer, but it failed to cook the meals. NPL got adverse publicity. Krishnan shut down the project and desisted from taking any more such projects.[347] It transpired that

> *"'the scientists of the national physical laboratory had designed the said cooker, inspired by a great thought that the then president had during his satyagraha days".[348]*

This example, of scientists taking up projects without critically examining, merely on the half-baked knowledge of the politician and later by science administrators in power is an example that was replicated at many institutions.

Appointing a scientist specialist in his field, to oversee the work of an institute whose aims and objectives had nothing to do with his core expertise results in the director using the resources of the institute to continue working in his own field of interest. This results in ambiguity of aims and the dilution of the institute's resources and its core competence. These scientist-administrator usually have their own group of workers who get preferential treatment. This often becomes a cause of heartburn to other scientists. Krishnan's group at NPL was no exception and senior scientists resented undue claim by his group members over Krishnan's time.[349]

The problem of square peg in a round hole did not remain confined to early appointments after independence. The hierarchical administrative structure for research institutes, also need to accommodate the scientists who attain seniority. Often a scientist with expertise in one field may be asked to take charge of a department totally unrelated to his expertise. A nuclear physicist may be sent to head electronics, or spectroscopy department or a theoretical physicist may be sent to a technology department and vice versa. An instrumentalist may be asked to head a biology group or sent to head a library or purchase department. Apart from underutilising the capabilities of the scientist, inexperience in the new area results in inefficient management. In addition, this disheartens the superseded scientist who is an expert in the field and has toiled over years for fulfilling the objectives of his group/department. This problem, endemic to science administration, remains unaddressed in Indian research institutions.

8.3 First National Awards To Scientists

In 1954 the government instituted national awards for achievements by individuals to recognise their contributions in their respective fields. The first awards and the selection of awardees brought to the fore the fact that excellence in one's field alone does not guarantee the national award.[350]

"In January 1954, the Government had instituted a scheme of national awards. .... The first set of awards was announced on 15 August 1954 and to great joy of the scientific community, the Bharat Ratna award was given to Sir C. V. Raman, the Padma Vibhushan Pahela Varg to Satyendranath Bose, While Krishnan along with Bhabha, Bhatnagar and J. C. Ghosh was awarded the Padma Vibhushan Dusra Verg. Missing from the list was Meghnad Saha, whose competence and contribution as a scientist was no less than that of any of the awardees above. But his acerbic criticism of the policies of the Government on the floor of the Indian Parliament had alienated him from the ruling establishment, and the Government failed to rise above its petty-mindedness to decorate him with the honour that he rightly deserved".

The political class, particularly Pt. Nehru, with his western education and commitment to the democratic principles miserably failed in setting up a healthy tradition, namely that of judging scientists by their contributions to science alone and not by any other extraneous factors like political belief, personal likes, dislikes and disagreements etc.

Saha was often critical of the policies of the government inside the parliament through incisive questioning and outside by writing well researched articles on various problems in Science and Culture.[351] Criticism of government's policies, even legitimate, had no room for expression in the society and was not taken lightly. Anderson in this connection writes;[352]

"On the last day of 1954 Saha wrote to Nehru about a statement Nehru made ten days earlier criticizing Saha's grasp of national income statistics, per capita income, and particularly of whether or not industrial production had risen 18 percent or 13 percent. This arose in the context of a debate in which Nehru described "my kind of science". Saha much resented the suggestion that he had not done his homework and wrote to Nehru about their entire working

relationship going back to 1936,..... Finally he closed this letter by evaluating the two people on whom Nehru most relied in atomic energy matters.....Saha's letter was written on New Year's evening, the very evening that Shanti Bhatnagar died. Showing their relationship was not casual, Nehru wrote promptly two days later to Saha about their disagreement over industrial production statistics. In his reply Nehru explained the path that the recent negative appraisal of Saha had taken to reach Nehru........ Then Nehru gave Saha his own appraisal':

> *After making a strong attack on everything that Government has done and running it down, you were good enough to compare us to Chiang Kai-shek and his failure. It seemed to me that your criticism was not only unjustified but completely lacking in objectivity and therefore most unscientific. You were evidently angry and lost sense of perspective. If you attack the Government; surely you do not expect them to remain silent. I can hardly judge myself. It may be that you are a better judge of me than myself."*

It is no surprise that Saha was not considered for the national awards. This ungracious act of overlooking Saha, by the first Prime Minister of the largest democracy towards an eminent scientist set the trend for future politicians and future scientists also got their cue.

It is more shocking that none of the scientists of the time protested. Of the five awardees scientist, Bhatnagar, Krishnan and Bhabha were the government scientists, supporting the policies of the government. Raman was not associated with government science, and was at times critical of it but, being a Nobel Laurette, he was too big to be ignored. The awardees scientists also failed to rise above their petty interests and feuds and did not protest against the government's deliberate intention of disregarding Saha's contribution to science. Least they could do was to decline the offer stating their reasons.

Pt. Nehru also could not rise above his antipathy towards Saha and disregarded Saha's scientific contribution and his international reputation. The writing was clear. Anyone opposing the policies will not be honoured by the government even if his scientific achievements are recognised internationally.

This resulted in establishing an eco-system where scientists now see the writing on the walls of the parliament clearly. All efforts are made by them to be on the right side of the government, nurture political connections, and showcase their individual achievements by sending hundreds of the copies of their biodata, including the list of publications, bloated by the contribution of junior colleagues, in most cases, for canvasing support for the awards. Independent unbiased critical opinion from the scientists to the government became a casualty.

8.4 'Blue-Eyed Boy' Syndrome

In earlier chapters we have seen how leading personalities, who were the office bearers of various scientific societies, could not burry their differences and amalgamate their societies into one in spite of the government's best intentions and efforts. The proposed unification was aimed at providing unbiased opinion on scientific matters to the government. The government in desperation depended on few individuals, like S. S. Bhatnagar, K. S. Krishnan and H. J. Bhabha to carry forward its agenda. Proximity of these individuals to the centre of power and the importance given to them created an unequal standing in the scientific community vis a vis other equally eminent scientists of the time. Prime Minister Nehru's over dependence on Bhabha, for instance, particularly after the death of Bhatnagar and Krishnan, placed Bhabha in an envious and privileged position. Bhabha was the only scientist who had direct access to the prime minister. He soon began addressing the Indian Prime Minister, not by usual protocol that the prime minister's chair deserved, but as Bhai (brother).[353] This uniquely overtly expressed relationship placed Bhabha in a singularly advantageous position.

How does a 'Blue-eyed Boy' come into being, In any organisation, increased administrative workload compels an administrator to find a suitable colleague who can fill in for him in day to day running of the institutes under his charge and provide timely feedback from the ground on the happenings within and act as a sounding box for his intended decisions. This is how a 'Blue-eyed Boy' syndrome sets in. In large organisations where most senior scientists are busy in attending numerous meetings, organised for a host of inane committees, this syndrome percolated down, creating an unhealthy environment at every level. Scientific research requires free and frank interaction between the scientists which, in the presence of 'Blue-eyed boy' is tempered, inhibited or convoluted and introduces a culture of 'palace intrigue' to the detriment of the eco-system.

A former BARC scientist, has eloquently described the making of a 'Blue eyed Boy'.[354] The incidence narrated there-in brings to the fore the usefulness of such 'Blue eyed boys' to browbeat or humiliate one's contemporaries and colleagues. This is a demoralizing trend for any institution and sets up an unhealthy working environment.

*

PART III

Social, Religious and Political Causes

9

Social, Religious and Political Causes

It is known that the Indian and Chinese civilisations were well ahead in science and technology in ancient times and yet modern science developed in the western world. Why? This question has occupied the minds of many scholars, including Needham who has extensively studied Chinese science.[355] Are there any cultural, social or religious factors that control the society's inclination to accept science? What role religion historically has played in shaping human mind and the society? It will be instructive to briefly look at the chronological development of religious, political and scientific events over the period of our recorded history and try to understand their interaction to look for answers for the present state of science in India.

Religion and science, both, seeking knowledge about the unknown, have existed together from ancient times.[356] The earliest evidence of some form of religious activity along with accurate drawing of the layout of the city itself is found in the ruins of Catal Hüyük Turkey, about 6700 BC - 5650 BC. [357]

> "A wall painting from one of the shrines, dated to 6200 BC with an error margin of less than 100 years, depicts the plan of the city and corresponds very accurately to the arrangement of the houses as they were excavated. It testifies for the advanced state of accurate measurement and mapping."

The coexistence of religious faiths and scientific enquiry about nature have had moments of distrust and a fight for supremacy throughout history.

The timeline of the history of the human activity can be grossly divided into three periods, namely the Ancient, Middle or Medieval Age and the Modern Age. The time period from about 3rd millennium up to about 5th century BC is generally considered as the Ancient age. In this era science developed in several civilisations around the world as is evident from the archaeological findings. The civilisations existing during this time were the Indian[358], Chinese[359] Roman[360] and Greek[361]. The scientific activities in these civilisations were coterminous and evidence of communications between the Indian and Greece exist. The period saw contributions to human knowledge from Pythagoras (570 – 495 BC), Socrates (470 – 399 BC), Plato (428/427 or 424/423 – 348/347 BC), Aristotle (384–322 BC), Claudius Ptolemy (AD 100 – AD 170). The most influential of them all, Aristotle's teachings and his geocentric model of the world, even though wrong, dominated the scientific and cultural discourse for almost 2000 years. A number of philosophers, holding views contrary to prevalent thought, were persecuted throughout history by the religious authorities/rulers and condemned to death.

Greek religion is not the same as Greek mythology, which is concerned with traditional tales, though the two are closely interlinked. Greek religion in its developed form lasted more than a thousand years, from the time of Homer (probably 9th or 8th century BCE) to the reign of the Emperor Julian (4th century CE). During that period its influence spread as far west as Spain, east to the Indus river, and throughout the Mediterranean world. Its effect was most marked on the Romans, who identified their deities with those of the Greeks.[362]

> "*'The period from 3000 BC to about 1000 BC known as Bronze Age was also synonymous with progress in astronomy, mathematics and the calendar. .... A qualitative change occurred in Greece shortly after 1000BC. The emerging Greek civilisation broke the link between religion and Science and established a new discipline for systematic attempts to understand and explain nature. The discipline was called philosophy, literally "love of wisdom", and science was part of it ....*
> *.........It has often been said that the Greek religion was not suitable as a tool to explain the natural world because it lacked the mystical depth and all-encompassing concept of other religions".*[363]

Alexander's father appointed Aristotle as his teacher when he was 13 years of age. In 336 BC Alexander was appointed as the king of Macedonia.... After his death in 323 BC Alexander's empire was divided amongst his generals and Alexander's bodyguard PtolemyI became the ruler of Egypt with capital, Alexandria, becoming the new centre for science. Ptolemy founded a museum and to build the base for a new university he decreed that all visitors to Alexandria submit the manuscripts in their possession for copying. The museum became the centre of all European knowledge and many Greek scientists spent time in Alexandria {Archimedes, Euclid, Hipparchus, Hero}. Alexandria flourished for

300 years under Ptolemies until Rome took control in 30 BC. The museum was severally damaged during the conflict.[364]

Roman civilisation grew but was characterised by encyclopaedic work. Roman society showed little interest in science but tolerated the activities of scientists. Greek science continued in colonial Egypt.[365] Roman Empire was consolidated out of city states and was always beset with conflict and prone to attacks by German tribal invaders. It was split into eastern and western part; the western part was later called as Byzantine Empire.[366]

> *"An internal force of unrest during the period of empire's decline was the arrival of Christianity, though some authors attribute it to disruption caused by the Barbarians. Christianity began as a religion of poor and oppressed, went through cycle of violent suppression, persecution, toleration but ended finally in 313AD with the Edict of Milan. Emperor Constantine was the first ruler to embrace Christianity. The end of western Roman Empire was the beginning of a dark period for science in Europe which lasted for an entire millennium.......*
>
> *Science was an activity of the ruling class whereas the Christianity had begun as a religion of the exploited and oppressed and as a consequence had no scientific tradition of its own. Its preachers spoke of miracles, and mystical powers and promoted a way of thinking that was diametrically opposed to rational scientific thought. They relied on scriptures which could not compete with writings of Greek philosophers, so they tried to recast Greek thought in to Christian doctrine. By the end of 4th century Christianity had adopted Greek philosophy but had no place for Greek science and could not tolerate the teaching institutions. In 391AD Christian mobs attacked the Serapeum of Alexandria, in which empress Cleopatra had housed the remains of its famous library, and burnt it to the ground......The end of Alexandria's Museum came in 415AD...... a mob of monks........and Alexandria fell to the rule of the church".*

What was the relation between Christianity and science?

"Christianity absorbed Greek Philosophy and adopted it for its own use. Rational science was a threat to Christian dogma and the church actively sought to eliminate it". [367]

The Roman Empire was built on slave labour and confrontation with the barbarian peasants led to the establishment of a feudal society where ownership of the land was with few and the peasants shared the produce with the landowner. In due course of time Church became landlord and held important political power. The Roman empire (Byzantine) collapsed in 1453 with the fall of Constantinople and the influence of the Catholic church waned. The Catholic clergy became rich through various taxes and selling of indulgences. The indulgences were the sins perpetuated by the individual and by paying to the church an individual could get rid of the sin and attain salvation. The corruption became so endemic to the society that these indulgences were freely traded.

Towards the end of the Middle Ages contact with Arabian science led to the establishment of several universities in Europe.

"An increasing stream of Arabic scientific literature arrived in Christian Europe. As a result, a great wave of university foundations swept through the continent. The university scholars spent their time translating and copying the Arabic translations of the Greek classical science texts but did not develop novel concepts or ideas. The great misery of the century following the Black Death stifled the universities as well. Nevertheless, their existence provided a good starting position for the development of science during the Renaissance that followed."[368]

The renaissance saw the emergence of new thinking in the fields of painting, sculpture, architecture and science. This period was preceded by the hundred-year war between England and France. The first European Renaissance of the 12th century was characterised by extensive translations of the Greek and Arabic works.[369] The second Renaissance spread over 14th to 17th century, [Italian (1300-1640), German (1400-1543), English (1515-1600) French (1494-1610)] was a cultural movement which started in Italy and spread to countries as far away as Russia. The Renaissance provided the backdrop for profound social changes that helped in spreading of science in western countries. Some historians consider the start of this period to be in 1620's. The publication in 1543 CE of Copernicus's article, (*On the Revolutions of the Heavenly Spheres*) a radical thought at that time, is considered as the start of the scientific revolution. This is the start of the Modern age.

In 1534 Henry VIII had separated the Church of England from the Roman church. A group of Protestants, called 'Puritans' were not happy with the pace of the reforms in the Church of England. They believed that not enough is being done to rid the Church of Catholic practices and to reform the structure of any liturgy, ceremony, or practices which were not found in Scripture. What was significant in puritanism was their emphasis on frugality, Industry and acceptance of materialism to achieve the ultimate goal.[370] Martin Luther (1483-1545), [371] a German theologian wanted to reform the Protestant Church. In 1517 he posted his "95 thesis". He was convinced that[372]

> *"Salvation cannot be earned or bought but is a divine gift; God's mercy is seen in Jesus Christ and the saved soul can serve God without the feeling of guilt"*

This led to the establishment of Lutheran Church.

> *"The establishment of the Lutheran Church was the first step for Christianity to adopt to the new capitalist ethic. ....Luther's contemporary, John Calvin, a purist himself, went a step further by declaring that hard work pleases God and therefore leads to wealth. Personal prosperity thus turned into a sign of a godly life. This form of Calvinism became the main Christian faith in the Netherlands, Scotland and England from where it spread to north America.'".*
> 373

Puritan exodus in 1641 to New England (America), due to persecution, was the ideological seed that shaped America.[374] Benjamin Franklin, the founding father of United States of America was a puritan and his views are reflected in the declaration of Independence. It subsequently affected the work ethics of United States of America (USA) and Canada. The Puritans believed in the separation of church and the state, but not a separation of the state from God.[375]

The renaissance in Europe was followed by a series of revolutions which saw a diminished role of religion in the state. English revolution or English civil war (1642-1660) abolished the idea of divine right of the king and established the supremacy of the parliament over the crown and ultimately limited its role in governance.[376] The French Revolution (1789-1799), after various campaign between 1793 to 1794 resulted, not only in public reclamation of the Church's properties and money but also the termination of Catholic religious practice and of the religion itself.[377] The period between 1715 to 1789 is considered as the Age of Enlightenment or the Age of Reason where intellectual and philosophical movements dominated the world of ideas and undermined the authority of the monarchy and the Church.[378] The agreement between Napoleon and Pope in 1801 restoring the role of the church in the society was terminated by law in 1905 and the church was separated from the state.[379] Russian Revolution of 1917 resulted in overthrow of the imperial government and placed the Bolsheviks in power.[380] The communist regime officially promoted atheism and the churches were closed and religious activities were suppressed.[381]

The Scientific Revolution took place in Europe towards the end of the Renaissance period and continued through the late 18th century, influencing the intellectual social movement known as the Enlightenment. Age of Enlightenment that followed Scientific Revolution gave way to the "Age of Reflection." (1800–1840). This was an intellectual movement that originated in Western Europe as a counter to the late 18th century enlightenment. Romanticism incorporated many fields of study, including politics, arts and humanities, but it also greatly influenced the 19th century science.

The progress of science through ages and the negative impact of Christianity (Catholic) is graphically summed up in figure [3]. [382] The hole created in scientific advancement by Christianity is beautifully illustrated. What the world would have attained in the year 1000 could be attained only in the year 2000.

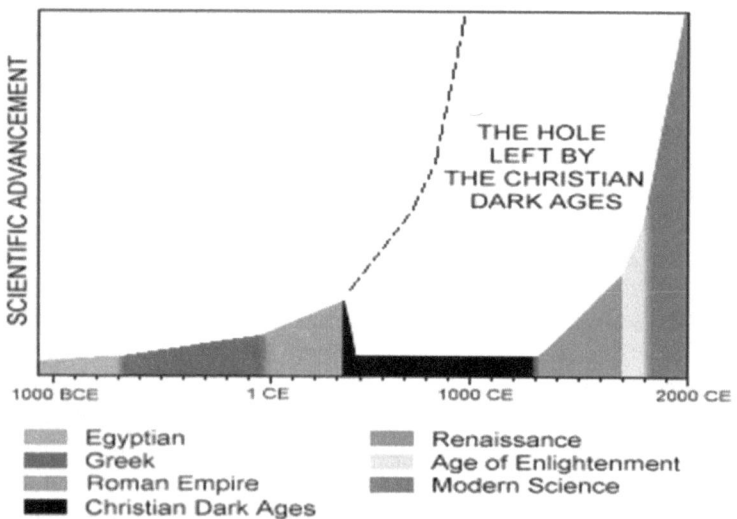

Figure 3 The Hole left by the Christian Dark Ages

Let us now see how science progressed over the centuries of mankind's recorded history. In order to study the effect of various factors on the growth of science at a particular time and the place we have to set up some parameters to monitor. With the complexity of interactions between the religious, political, social and geographical factors with emerging new science in any society it is difficult to define a common parameter. In order to quantify the growth of science, we propose to consider the active period of the life span of a scientist as one parameter. The number of scientists in a country at a particular time can be considered as an indicator of the growth of science in that particular country. The data about the scientists and their life span is taken from various sources on the internet, knowing fully well the unverifiable and unauthenticated nature of the information. This can, however, give relative measure of the advancement of countries in science. The year of birth or the year of death is chosen to link the contribution of a scientist to the century of his existence. Assuming, the lifespan of a scientist to be generally less than a hundred years, as is mostly the case, this assumption seems to be justified. Some ambiguity might creep in the active period of a scientist if the birth year falls at the beginning or in the closing years of the

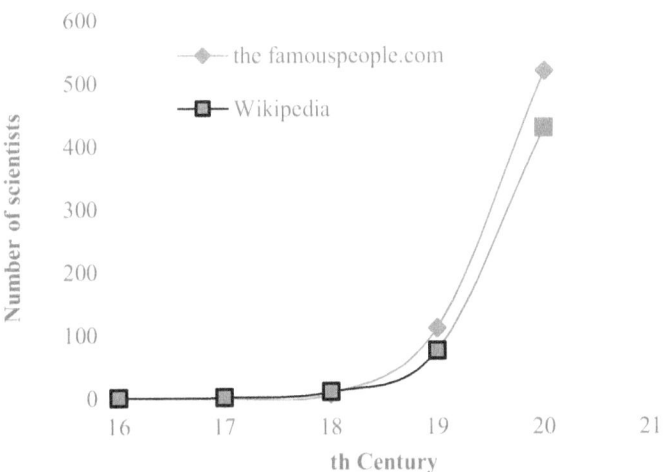

Figure 4 Comparison of data from two internet sites for American scientists

century. The active life span in these cases covering the transition years would not make much difference to scientists' inclusion in either the previous or post birth century. In order to check the compatibility of this approach, the data about the American scientists from the two internet sites was checked for the same fields of activity [Fig. 4]. The trends of the two data are similar, though absolute numbers do differ.

Using this method, we will now consider the growth of science in various countries and try to understand how social, religious and political factors affected it.

Figure [5] gives a timeline of the growth of European science, represented by the number of scientists in pre-Christian and post-Christian era along with the changes in the Christian theology.

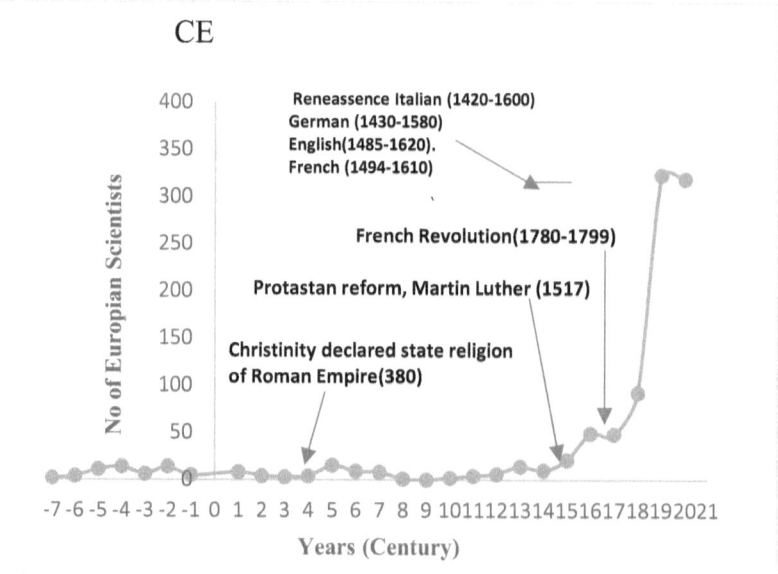

Figure 5 Growth of European science and important religious and social changes

After the declaration of Christianity as the state religion of the Roman Empire (380 C) it spread to a large area of Europe. The dark ages of science, thereafter, lasted for about 1000 years until the Renaissance in 1420 C. Scientific activity began to pick up, but the real fillip, however, came about when in 1517, a Protestant preacher Martin Luther, called for the reform of the corrupt practices of the

Catholic Church. Martin Luther and after his death another preacher Calvin introduced material gain also as a worship of God.

Religion is an antithesis to Science can, more clearly, be established by taking example of the Islamic science whose rise and decline both were dramatic. The period in between the Ancient and the modern age, the Medieval age, saw the golden era of the Arabic Science (800BC -1100BC). Arabic science or the Islamic science became a precursor to the growth of western science in almost all its branches.[383]

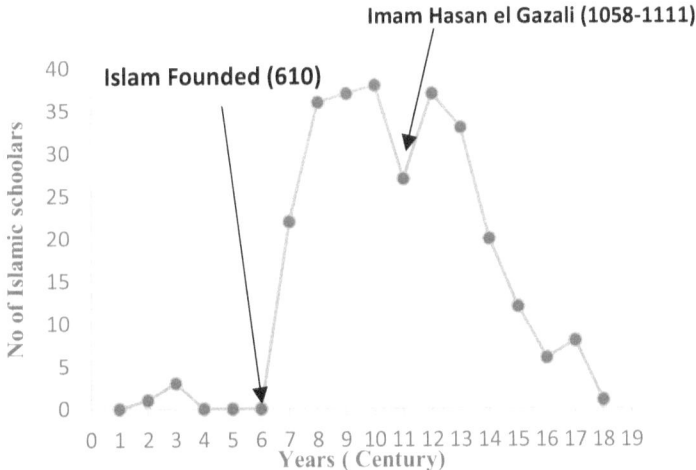

Figure 6 Rise and fall of Islamic science

Figure [6] shows a timeline of the number of Islamic scholars mainly astronomers, mathematicians and medical practitioners, who contributed to the growth of Islamic science.[384] Science rapidly grew after the establishment of Islam in 610 CE. Islamic science did not grow in vacuum. It was supported by two major dynasties, the Sassanian Empire (224-651) and the Byzantine Empire (286-1453).[385]Arabic science absorbed the scientific knowledge from Indians[386], Chinese[387], Greek[388]and Christians.[389] The exchanges and sharing of knowledge occurred through extensive translations,[390] account of travellers, and interaction with merchants. This golden era of Islamic science however, ended and all the gains of science were subsequently lost.

Why did science flourish in the Islamic world with such intensity and why did it decline after 1291 CE, never to recover? The two factors that were responsible for its rise were, high importance to learning in Islam and its cradle being located at Baghdad. Baghdad was at the confluence of different streams of cultures Roman, Greek and regions of west Asia, India and China.

The Islamic society was open and receptive to new influences. Christian scholars translated the works from Greece to Syriac and then to Arabic.[391] It should also be noted that the rulers of the Abbasid dynasty (750-1258) emphasised the importance of science, philosophy, medicine and education.[392] This benefited Islamic science in its ascendance.

The start of the decline of Islamic science is attributed by some to Imam Hamid el Gazali (1058CE-1111CE),[393] a preacher who is believed to be a 'mujaddid', a renewer of the faith who appears once every century to restore the faith of the Islamic Community. Classical Sufi scholars have defined Tasawwuf as;[394]

"a science whose objective is the reparation of the heart and turning it away from all else but God".

Gazali was a learned and respected theologian, a preacher and a prolific writer. His disdain for mathematics is well documented.[395] Thus, Gazali, through his writings and preachings, set the clock back to 'revelation' replacing 'investigation' of the golden era of Arabic science.[396] Some attribute the decline of Islamic science to the war of crusades (1096 -1291)[397] and plundering of Baghdad by Mongols in 1258.[398]

It is thus clear that religion plays an important role in growth or otherwise of science in ancient and medieval times. Let us now see how modern nation states have fared in spread of science. Figure [7] gives the number of scientists from the data gleaned from the internet for a number of countries. One site, thefamouspeople.com[399] did not have sufficient data about the Russian and Chinese scientists.

This data was gleaned from Wikipedia.[400] How is it that the American growth of science is not constrained by religion and its manifestations in the society? How could the American society, composed mostly of the European settlers, free themselves from the negative influences of religion? Is it that the American society had burnt the bridges after crossing the Atlantic? Was it a new social structure that was set in place, a social structure that equated acquiring wealth as means of worship? The renaissance in Europe and the revolutions that followed did marginalise the religious order, but did it free the society from old traditions, rituals and beliefs?

Russia was ruled by monarchy under the orthodox Christian church from 862 to 1918. In Russia science, given some importance by the Tsar, had established some tradition of scientific research before the revolution.[401] The revolution in 1917 brought in the Communists to power who banned religion and confiscated church properties. The growth of science in Russia was phenomenal particularly after the revolution. The breakup of USSR in 1991, resulted in drastic reduction of state funding for scientific research and migration of scientists to other

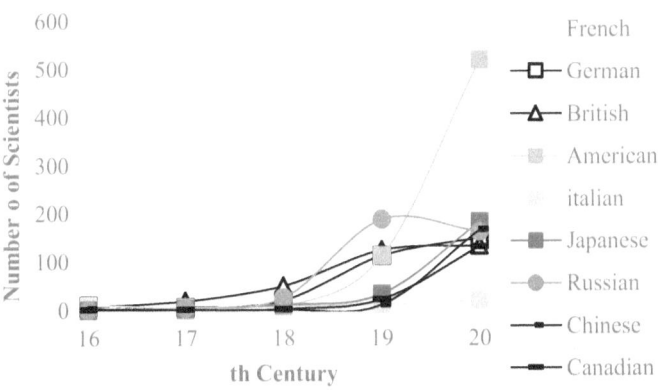

Figure 7 Growth of science in advanced countries

countries. That led to a drastic decline in its growth rates. The growth of the scientific enterprise is half of what it was before the breakup.[402]

Japan's Meji restoration [403] was a social change that paved the way for resurgence of Japanese science. The two important oaths, amongst others, that the Japanese society took, were;

(1) evil customs of the past will be broken and
(2) knowledge shall be sought throughout the world. [404]

These oaths, coupled with the agreement for transfer of technology with the Americans, helped Japan to become world leaders in technology.

The Chinese civilization began as slaveholder society and continued as feudal empire from its unification around 250 BC to 1949.[405] In 1947, China was beset with civil war which ended in Maoists taking control of the government and establishing the Peoples Republic of China in 1949. From 1949 onwards a series of events, the long march, cultural revolution (1966-1976) etc changed the traditional Chinese society beyond recognition.[406] The social structure was destroyed at considerable hardship to the population and the religion was pushed to the background. Chinese science was yet to emerge on the world stage.

It would seem from the above data that, apart from the negative impact of religion on the growth of science, there is a need for a change in the society for providing an atmosphere conducive for science. An established social structure under centuries of religious dominance carries with it the under currents of regression. A sharp rise in the growth of science is observed in the countries that changed the social order in addition to marginalizing of the religion.

It is clear that religion plays an important role in assimilation of science in the society. Its negative effect on the development of science can be seen throughout history. The societies that have progressed have relegated religion to the personal domain and reduced the importance of rituals to a minimum. Generating wealth is not demeaning in these societies. Renouncing worldly gains is not glorified.

Russia and China went through the social upheavals to attain position of prominence in science and technology. It is, however, not always necessary to disrupt the society for progress of science. However, examples of America, Japan and Canada do indicate that pushing religion to a secondary role makes the progress of science in a society more effective and meaningful.

Let us now explore as to what form of government is most suited for growth of science. The world has experienced various forms of governments from monarchies, totalitarian regimes and democracies with varying degrees of individual freedoms, corruption and the enforcement of law. Which parameters are best suited for growth of science?

Since scientific progress of a country can be linked to the number of technological products and processes patented, year wise number of patents awarded to a country can be a good indicator of the technological progress of the country. Figure [8] represents the data on the year wise number of patents held by a country gleaned from internet.[407]

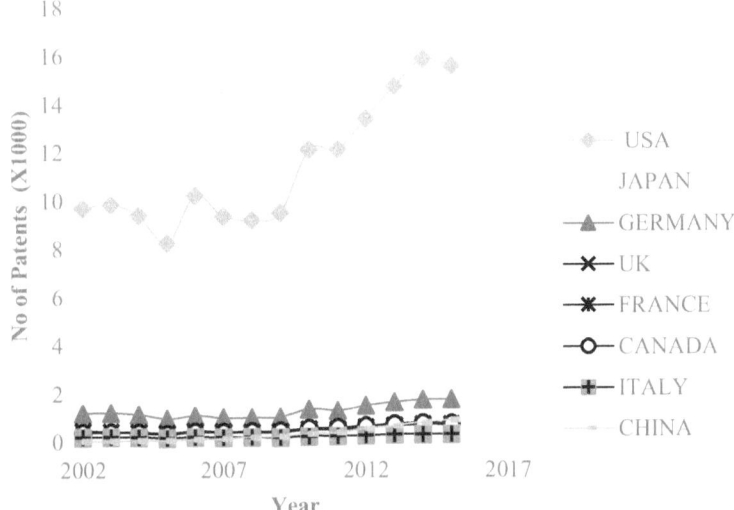

Figure 8 Country wise growth of Patents

From Year 2002 to 2016 highest number of patents are held by United States followed by Japan and Germany. Russia and China are way behind. Considering that technologically Russia is very advanced, the number of patents to its credit are significantly less than those for western democracies. Loren Graham in an interview to Globe correspondent points out that there are hardly any innovative products originating from Russia that have made an impact in the world. [408] He discusses this discrepancy and concludes that for science to progress and the society to reap the benefits of science it should be free of corruption.

Graham's comments on the organisation of Soviet science seem to be pertinent for India. [409]

"Soviet science and technology were highly centralised and ruled from above. In fundamental science, the Academy of Sciences, a network of several hundred institutes, acted as the major organizational force. The academy's budget came from the central government and was distributed from the top down, in block grants. There were no fellowships and grants for which individual researcher could apply, and there was no system of peer review. This mode of management and financing research gave institute directors enormous powers, for they controlled the budget. Senior researchers with influence were much more successful in garnering research funds than were junior scientists. Further, research and teaching were largely separated, with the academy responsibility for research and the universities assigned a primary pedagogical role. The result was an exceptionally elitist organization of science, a system in which less prestigious researchers, such as the young or university teachers had great difficulty in fulfilling their potential. Nonetheless, this system worked fairly well in cases where the elite scientists in charge were talented and productive, as was often the case in such fields as theoretical physics and mathematics. But it worked poorly when second-rate scientists ruled, ……".

The corruption inhibits an intellectual mind to go into business. The Graham's views are supported by the corruption data. [Figure 9] The corruption index for the technologically advanced western democracies hovers between 20 to 30. In contrast Russia, China and India fall in the range of 60-75.[410]

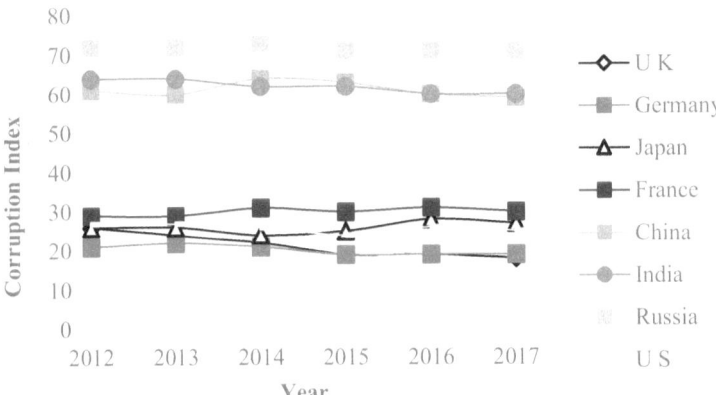

Figure 9 Comparison of Corruption Index of some countries

Figure [10] shows country-wise number of patent applications for from 1962 to 2017.[411] Dramatic rise in number of patent applications from China surpassing even USA in last decade, a communist country with capitalist economic model, probably offers a different scenario for the growth of science, an exception that defies Graham's prognosis. China now stands next to United States in terms of number of publications, has second largest number of eminent scientists and has made significant advances in all areas of science and technology.[412]

China defies another axiom that the high level of freedom of expression and respect for contrary views in western democracies led to high achievements in science as compared to Russia. The censorship index for China and Russia at 80 is much higher than that for the western democracies, that fall in the range of 17 to 25. India, with index of 37, falls in between the two extremes.[413]

It can thus be summarised that relegation of religion to background, democratic form of government, low level of corruption and high index for freedom of expression are facilitating factors that have helped western countries to make advances in science. China has relegated religion to a secondary role political structure that controls every aspect of the society. It has complete control over the economy, society and the natural resources.

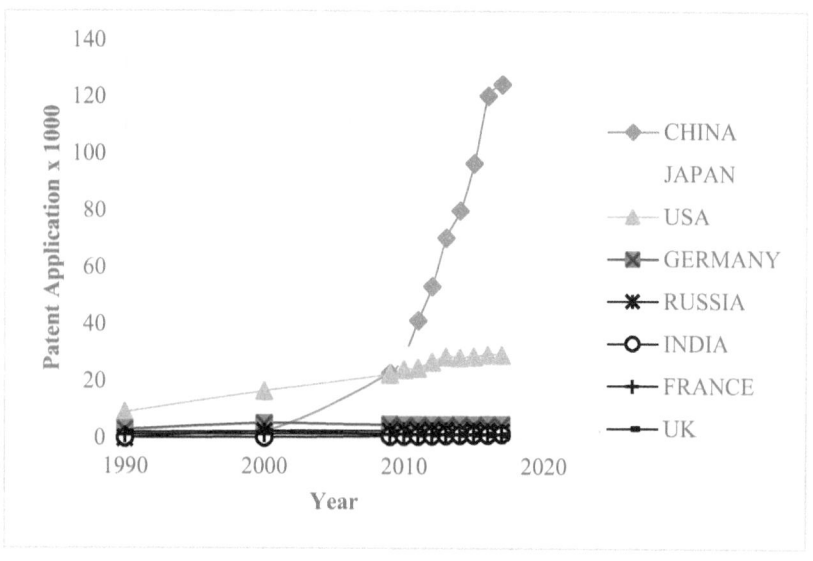

Figure 10 Growth of Patent applications

Let us now see how religion has influenced growth of science in India. Figure [11] shows the growth in numbers of the eminent scientists[414] and prominent Hindu religious gurus[415] over centuries of Indian civilisation. The see-saw manner in which religion and science vie for dominance, particularly during the period of foreign rule, is clear. The period of occupation by the Muslim invaders and the British saw a host of religious sects and movements. The population sought solace under the canopy of myths, rituals at the feet of saints. The parallel rise of the two contrary precepts, in the society, is surprising.

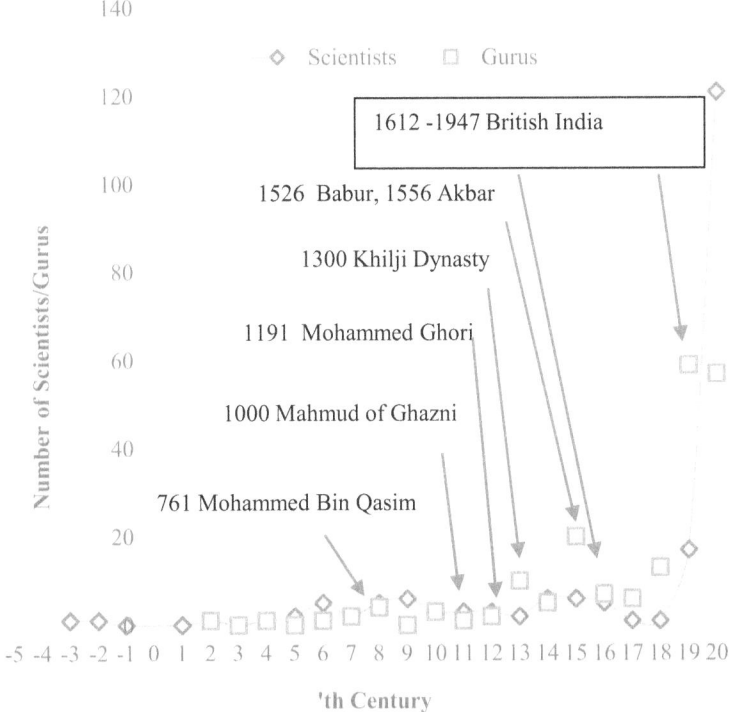

Figure 11 Growth in number of Scientists and Hindu religious gurus during foreign rule

Figure [12] shows growth of India's population[416] and the attendance at one of the important Hindu religious festival, the 'Kumbha' since year 1900.[417] Significant increase in Kumbh pilgrims in recent times indicates increased role of religion in the society.

The percentage of Indian population attending Kumbha Mela shows a decline in the beginning of the twentieth century but steadily increases after independence and shows a sharp rise between the two Kumbha events of 1989 and 2001. The number of pilgrims increased from about 2.2% in '89 Kumbha to about

9.1% in 2001. Even after taking into consideration, liberalization of Indian economy and increase in average prosperity, the increase in influence of religion over the society is remarkable. Did infusion of societal scientific temper take a hit?

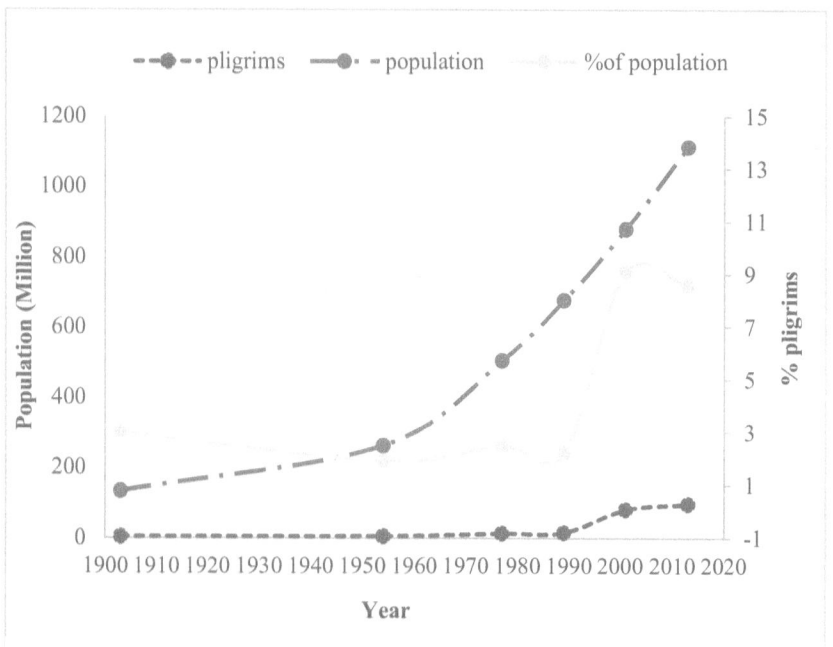

Figure 12 Growth of Indian population and the number of 'Kumbha' pilgrims

Western science was often seen as an afront to the Indian culture which came to be limited to rituals and religious discourse by increasing number of saints, sadhus and preachers. Orientalism, that placed the spirituality and the glorious past of the Indian civilization on a high pedestal, gave a sense of superiority over the materialism of the west giving birth to a host of modern-day post-independence religious preachers like Rajneesh,[418]Mahesh yogi[419] and others acquiring wealth and fame. Rise of western science was driven by pursuit of wealth which was declared as a form of worship. Renouncing wealth and worldly comforts have always been regarded as supreme sacrifice and the ascetics are treated by the society with extreme reverence. Bose's refusal to patent his invention and reap benefits for himself made him a 'rishi', an ascetic. This probably did more harm to Indian science.[420]

Let us at this stage revisit Basalla's model and concentrate on its third phase in the Indian context. The colonial powers have left the shores and the country, and the society is left to itself. Basalla points out, in the context of America, that immediately after the removal of foreign rule nationalism takes over. [421]

"...after the American revolution there was nationalistic sentiment in the new nation which encouraged the building of an American science upon a native foundation".

Pre-independence Indian scientists were motivated by nationalistic zeal. After independence, nationalism did take over but what was the resolve of the Indian scientific community? The eminent scientists were not united in the way the science should progress. The multitude of scientific societies were fighting for the turf and hegemony in spite of the government's pious intentions of setting up of some mechanism for much needed unbiased advice from the community of scientists. Ambitious individuals busied themselves in manoeuvring a space for their own ambitions without regard to the damage they were to inflict on the fragile educational system and the ethos of the scientific research.

Basalla further writes [422]

"If a colonial, dependent scientific culture is to be exchanged for an independent one, many tasks must be completed. ....as follows."

1.   *"Resistance to science on the basis of philosophical and religious beliefs must be overcome and replaced by positive encouragement of scientific research. Such resistance might be ignored or circumvented by the colonial scientists, but it must be eradicated when science seeks a broad base of support at home.........".*

Basalla's above comments can be seen in the light of what Kapila says about the difference between the rise of science in Europe and that in India;[423]

"The emergence of science in Europe was an Event, in that it was a rupture in the pre-existing arrangements between knowledge, religion, and authority broadly construed as the Enlightenment tradition. The Event of science was not constituted simply by its ritualized contestations over disciplinary

exclusivity; rather, the specific eventuality of science in Europe was ultimately constituted by a confrontation between man and God. Whether this involved his "death" or his "exile," science had led, despite the dissenting tradition within the Enlightenment, to a categorical disenchantment with God. By contrast, in India science was no Event. .... The acceptance of science in India, in fact, defied the European terms of reference. Neither the exile nor the death of God could ever be declared—that is to say, it was never part of the possible. .... the argument here is that the work of science was to reformulate religion and to bring man back into converse with God, though on an entirely new footing. In other words, while the exile or death of God may not have been inevitable, even in the European world, the inevitability of science did not have the same political or religious consequences outside Europe and, specifically, in India...."

"........., religion in India did not emerge as a site for the reprieve or critique of science, nor was it that science was spiritualized. Rather, religion became the site of a disenchanted rationality."

Science and religion co-exist in the Indian society with religion having a higher status. One must at this stage differentiate between science and its end product technology. While technology is readily accepted and appreciated, presence of science is accepted with indifference. While technology has a visible presence, science on the other hand works behind the scene and the relation between the two is not obvious and the Indian public confuses technology to be science itself. The achievements of ISRO in space and work of DRDO in missile technology are considered scientific achievements even though the science and technology is established elsewhere. These are excellent achievements of technology and should be appreciated and be proud of as a nation. However, the reports in press of ISRO scientists performing pooja before a launch[424] or skipping inauspicious number 13 form PSLV launch number[425] reinforces belief in religious rituals and the occult practices and does little for the advancement of modern Indian science. In almost all offices, including the scientific research organisations, Vishwakarma puja is religiously performed every year.

2.

"The social role and place of the scientist need to be determined in order to insure society's approval of his labour. If science in general and, or some aspect of the scientist's work, is considered suitable only for the socially inferior, the growth of science may be inhabited. When Louis Agassiz visited Brazil in 1865, he was surprised to find that the higher

social classes held a strong prejudice against manual labour. This
prejudice had its effect upon the development of science in Brazil.... "[426]

Where does a scientist in India stand in the social order? The government
provides a very reasonable compensation for scientist's efforts as compared to the
average income. However, the society, including the scientists is divided on caste
lines. The caste structure creates a chasm between the artisan, who is of lower
strata and the high caste purveyor of knowledge. Manual work is often looked
down by the society, including the scientists themselves with the result that the
link between knowledge and skills so essential for innovations is broken. This
probably led to ancient Indian science lose its charm and still hampers modern
Indian science. In ancient China the craze for administrative jobs drew away
brilliant students causing its decline. The essence of Needham's argument is that
antagonism between manual and mental work runs in all civilisations. It is the
merchant class that brings together the manual and the mental work for progress
of science if it comes to power.[427] This is nowhere in sight presently in India.

3. *"The relationship between the science and the government should be*
   clarified so that, at most, science receives state financial aid and
   encouragement and at least, government maintains a neutral position in
   scientific matters...".[428]

This relationship between science and the government in India is not free
of the strings. The government funds the scientific research but also can direct
research in areas consistent with the ideological dispensations of the party in
power.

4. *"The teaching of science should be introduced into all levels of the*
   educational system....".[429]

Teaching of science is introduced at least up to the tenth grade. However,
the science teachers as well as the students, both are cut off from the class of people
who work with hands in different professions. Chasm between the practice of
science and the theory behind it leaves students to the confines of the textbooks
and ever-increasing number of guides. The system of evaluation based on ancient
rote method does little to ignite the minds of the students.

5. *"Native scientific organisation should be founded which are specifically*
   dedicated to promotion of science....".[430]

We have seen the growth of various scientific organisations before independence. These organisations and a multitude of associations that sprang up after independence do serve the needs of the scientific community by providing a collective bargaining platform in addition to providing a common ground for subject related exchange of ideas. Apart from organising national jamborees or mega meets, these organisations, do not address the problems faced by the working scientists.

6. *"Channels must be opened to facilitate formal national and international scientific communication. This can be accomplished by founding appropriate scientific journals and then gaining their widespread recognition.... The colonial scientist, who is accustomed to writing for established European scientific journals, may not wish to jeopardize his international reputation by reporting his work in an unknown native periodical......Finally, there are difficulties presented by language. Should national pride dictate that the contributions to the new journal be printed in the mother tongue when that language is not familiar to western Europeans, ... This was the question faced by the founders of scientific periodicals in Japan, and China, and in ......Central Europe. ...it is important that a country struggling to create an independent scientific tradition should publish journals of science filled with researches of its own scientists........".*[431]

A scientist in India prefers to publish his quality work in an international journal, giving reason for better exposure of his work and reserve his lower rank work for Indian journals, often to boost the list of publications. It is also a fact that most journals find it difficult to keep their publication schedule, for various reasons, preventing timely exposure to a scientist's work. The international ratings of most Indian journals are so low that the published work hardly gets noticed. The journals also need scientist-editors of high moral values. It is difficult to see the impact of Indian journals on the scientific traditions when editorial ethics are far from ideal. In Indian science administration, while assessing a scientist's work, lower weightage is given to publication in an Indian journal. This compels scientists to seek publications in foreign journals.

7. *"A proper technological base should be made available for growth of science. Western Europe had reached an advanced state of technical progress when modern science first made its appearance, and since that*

time it has been assumed that the two are fundamentally related. The exact nature of that relationship has not as yet been revealed by historians of science and technology. Even without clear guidelines it is possible to indicate some of the links between science and technology that are significant for this discussion. A nation hoping to be self-sufficient in the realm of science certainly must maintain a level of technology that will produce the scientific instruments and apparatus needed for research and teaching. ...... Fine scientific instruments, to be used by American scientists in research, teaching, and exploration, were customarily purchased in England and France until the second half of the 19th century. If America found it necessary to rely on Europe, one can imagine that an African or Asian culture, existing beyond the influence of Western technology, would find it much more difficult to reach the desired technological level and make its own instruments ...... ".[432]

This is an important aspect. India did not have sound technological base at the time of independence. It was not in a position to make scientific instruments so essential for making India self-sufficient. The economic situation did not allow for import of scientific instruments at free will. India could not have emulated America. Self-reliance became the official policy and most researchers busied themselves in re-inventing the wheel with shoddy claims of import substitution developing half-baked technology. Import substitution became a buzz word for a long time. When economic situation improved, import of sophisticated instruments flooded in, but years of languishing in back waters left the scientific community rudderless. Unlike Japan and China who absorbed the technology and improved it through reverse engineering and innovation, India failed. The growth of modern Indian science in turn suffered because of lack of advanced technology and instrumentation.

Basalla, places America and USSR in the most advanced first group, western Europe Japan Australia and Canada in second group and places India and China and some South American and African countries in the third group which has great potential. [433]

"China, India, and perhaps some South American and African countries may be placed in a third grouping of nations with great potential for future scientific growth and with major obstacles to be overcome before they establish their independent scientific cultures ..... ".

While China could overcome the constraints in the growth of science, India continues to battle with the constraints enumerated by Basalla. It seems that India has not been able to remove the obstacles and free itself from the colonial dependent scientific culture and has not yet been able to replace it with one of its own.

It is seen, from the analysis of the social factors [Part III] that felicitated absorption and growth of science in western countries. The structural changes in the society helped countries to create an environment that made them accept a rational outlook. The countries that changed the age-old social structure progressed at increased pace in science. India got an opportunity during the independence struggle that could have changed the social structure and also minimised the role of religion in the society. However, that was not to be. Mahatma Gandhi and other leaders of the freedom struggle used religion to unite the population to move together for freedom from the British rule. His efforts at eradicating untouchability and the constitution maker's efforts at restructuring with reservations for certain sections of the society did ameliorate the deprived section of the society but could not drastically change the social structure and hence could not establish a conducive environment for the science to find roots.

When the British left India, the social structure remained unperturbed even though the upheaval of partition did introduce some disruptive forces in the Indian society. India opted for a secular society with the Indian constitution carrying a resolve to promote scientific temper. Secular polity gave equal importance to all the religions. It did not negate or subdue the influence of religious dogmas and practices. Even though the state was not involved directly in promoting any single religion, in practice the government could not distance itself from the rituals of the Hindu tradition of performing puja while embarking on any new venture in its own institutions. The government, instead of disassociating itself from large religious gatherings of people, took responsibility of organising the events, under the cover of maintaining law and order, reinforcing the status-quo.

India opted for democracy and universal suffrage. With largely uneducated and deeply religious population the political class found ready means of influencing the voters and surreptitiously or brazenly touted the religious card. With the population kept engaged through religious events, rituals and discourses the scientific temper in public discourse was relegated to the background. The positive coefficient of exponential of the third stage of Basalla's model for the

Indian science was negatively impacted by all these factors, namely increased role of religion, increased corruption, and the Indian scientist's dilemma.

A scientist in India has a split personality. He is like a character in Hindu mythology,' a trishanku'[434] to borrow an idiom from Deepak Kumar.[435] He, is Baconian in the laboratory but submits himself unquestioningly at the feet of his deity and godmen. In a religiously loaded milieu, he forgets the Baconian principles, of applying inductive logic, experimental observation and re-verification before coming to the final conclusion,[436] that form the core of modern scientific enquiry and succumb to the temptations of offering obeisance to Godmen. If public sees eminent scientific personalities[437], bowing subserviently in front of godmen and the highest constitutional functionaries like Dr. Rajendra Prasad,[438] a brilliant science graduate, seeking solution to his worldly problems from his spiritual guru[439] common man's faith in science is shaken and the scientific temper gets a hit. Added to this is the notion of a glorious past that hampers clear and rational thinking. Every civilisation on earth has had its ancient past. One should be proud of one's own heritage but not be obsessed with it and certainly not find refuge in it. When politicians invoke the ancient texts to buttress the achievements of Indian civilisation, and denigrate modern science, stressing that every modern scientific law and invention was already achieved thousands of years earlier at premier gathering of scientists, the scientists themselves sense the direction of the wind and supple along.[440]

At the dawn of independence India woke up to a government structure that was left behind by the Imperial rulers. Centralisation of scientific and industrial research was on cards before independence by the then government. It was uncritically accepted and reinforced by the new government. Deepak Kumar in his article has eloquently summarised the situation,[441]

"Science administration in Victorian India mirrored colonial administrative policies in general - a top heavy structure as well as the existence of professional jealousies. Even after independence scientists took up administrative responsibilities, and committees became more important than classrooms. By this time, concepts of state science and state scientists had taken up a permanent place in the Indian system and psyche".

The eminent university researchers, who attained power and position in the state science setup and did reach high echelon of the scientific establishment, were captive of their antipathy towards universities. Bhatnagar blamed politics for the situation:[442]

"The universities in this country have not suffered for want of government help but the public interest in the universities has declined largely because the universities are having vice-chancellors not on the consideration of their attainments but of their political affinity. The net result is that the public hesitate to come forward and the government has to finance all the universities which it cannot do alone with its meagre resources".

Politics is the reason for decline of the universities after independence. The British advisers to Pt. Nehru had cautioned against depleting the universities of the talent but the Indian scientists thought otherwise[443]

"He had earlier criticized the fact that the CSIR pressured Sir K.S. Krishnan to leave basic research in a university, offering double his professor's salary, to ask him to try to build a research tradition applied to industrial problems. Blackett had seen this same phenomenon at Britain's National Physical Laboratory at Teddington. Both Blackett and A.V. Hill criticized the practice in India. D.S. Kothari also criticized Bhatnagar for taking scientists away from the universities, but his own Defence Science Organization did the same thing. Bhatnagar and Kothari argued that there was no other source of competent people."

And yet Bhatnagar was appointed as the first chairman of the University Grants Commission.

The building of educational institution also requires the grit and determination of a few who brave the odds. The illustrious examples of the vice-chancellors like Sir Ashutosh Mukherjee[444] of pre-independence era who fought all odds to put their institutions to academic excellence were forgotten in a short time in order to serve the self-interest of the new science administrators. It is not prudent to put the faults at the doorsteps of the universities when at that stage introspection was the need of the hour. While denuding the universities of talent, government ignored the fact that out of twenty-seven British Nobel laureates[445] in sciences, up to the time India gained independence, (1902-1947) almost all except one hailed from the British universities. The exception was a peripheral scientist, Rolland Ross[446] who worked for 25 years at a Calcutta hospital. The only native Indian to get the award, Sir Raman was also from an Indian university. It is inconceivable and unpardonable how Nehru, himself a product of western university, could not see that the universities and not in the government-controlled institutions were and continue to be the cradle of western scientific growth.

In an effort to improve the lot of education and a need to prepare engineer for the planned growth, a number of institutions of excellences like the IITs were opened. The requirement of teachers for these institutions deprived the universities and the British era colleges of capable individuals. These islands of excellence did produce engineers with high quality logical capabilities but a small percentage of the total graduates served the fields in which they gained knowledge. They preferred to go into management streams serving multinational corporations in foreign countries. The science and technology in general and the Indian science and technology in particular did not benefit.

Pre-independence universities were, by and large, staffed with eminent personalities. The trend continued even after independence for couple of decades. As local population got progressively educated, horizontal induction of teachers and research scholars from outside, declined due to employment demands for the 'son of soil'. This led to inbreeding and a decline in the teaching and research standards. The scientific research organisations, that were started with much fanfare, started feeling the effect of the marginalization of the universities. The deficient knowledge of new graduates forced them to start inhouse training programmes to cater to their demands for suitable manpower. The new organisations, started with highly accomplished individuals at the helm, failed to nurture the entrants. The lust for power did not leave talented administrator with spare time that is so crucial for mentoring a researcher. The administrators failed to establish healthy traditions by their actions and ethics within organisations suffered. The result was a growth of organisations devoid of healthy traditions and ethics.

Prof. Gill, compares the scientific ethos of the west and the counterpart in India;[447]

"An insight into the minds of Indian scientists who rose to power since independence throws some light on the scientific community in India today. Having been trained in a country where eminent men of science, weather in government or in the universities, were always ready to help one another and their juniors, I grew accustomed to certain protocols for working cooperatively. By and large, colleagues worked without suspicion that their

seniors or peers would take advantage of their contributions. A helping hand was always there when needed. With this background and idealism, I returned to India to give what I could to the country of my birth, especially since I had developed a strong nationalist sprit as a youth".

Commenting on the post-independence scientific structure, Gill compares the universities in India and those in US that take up work for Atomic Energy and Defence. He emphasizes that the universities should not be subordinated to the government institutions. The government and the industry can seek advice from the universities. Pure research is possible only in the universities where it is not bound by rigid rules and regulations. He advises that the university professors should have less teaching load so that they can spend more time guiding the research students.[448,449] *{Note-25}*

Gill also writes about the scientific administration,[450]

"'What we have created in place of a coherent system is a political universe with various categories of scientists in desperate orbit. The super-scientists are those who orbit closest to the nucleus of the government. Though few in number, they exercise a very powerful influence. They are know-alls and have-alls who have preserved their empires by entrenching themselves in all sort of committees, regardless of whether the decisions reached there pertain to their own work. ....They are not directly engaged in research but live on their past achievements".

"The second group consists of sycophants who occupy positions of importance in the scientific community. They have to remain in the good graces of the super-scientists in order to wield any power at all. The third group consists of the much-ignored university scientists who must please the first two groups in order to carry on any work. ......"

"The farthest orbit consists of scientific workers who find little or no encouragement. There are so many hurdles to overcome before they can get the equipment they need. If they desire to leave their place of work for a more congenial atmosphere, it is simply not allowed. Anyone who succeeds in leaving a laboratory where he is not happy runs the risk of continuous harassment. The final result is a deep frustration that leads to the easy solution

of going abroad. They do not go solely for money, but for the satisfaction that they wanted and appreciated".

"The famous geneticist drew attention in 1948 to two glaring defects in Indian Science: the eagerness of the leading personalities to remain in the forefront and their reluctance to give due credit to juniors working actively in the field".

It has become clear, after the experience of last seventy years that a fresh thinking is needed to rescue scientific research from the abases in which it presently is.

*

PART IV

The Crossroad

10

Where do we Stand?

We now stand at the crossroad more than six decades after the science policy resolution of 1958, drafted by H. J. Bhabha and passed by the Indian parliament.[451] Much water has flown through the confluence of the rivers Ganges, Yamuna and mythical Saraswati of the Vedic times *at Prayag*. The science policy resolution is now in government archives. The glory of our ancient science beacons us to bask in its afterglow while contemporaneous China, surges ahead in modern science at a speed unparalleled in the history of science. In the same time span as ours after independence, China is now in a position to challenge the western dominance in science and technology. The questions that now face us are, where do we stand? Is there a way out?

Where does a young aspiring scientist in India finds himself in the hierarchy of the multitude of science institutions? Looking around, the scenario is rather daunting. He comes across news reports of two senior scientists of the prestigious organisation, Bhabha Atomic Research Centre (BARC), that was so fondly and fastidiously crafted by Bhabha, coming to fist fight at their work place on accusation of plagiarism.[452] He sees an ex-Vice Chancellor being sent to jail for the same misdemeanour.[453] He comes across the news of a number of Vice-Chancellors being removed from the post on same charges.[454] He sees that it took three Nobel Laurette to write to the then president A. P. J. Abdul Kalam to take action against an erring vice-chancellor.[455] Leave aside the universities, even IIT's falling prey to infamy astonishes him.[456] He is aghast at the goings on at CSIR, so ambitiously founded and nurtured by Sir Bhatnagar.[457] He is at the end of his wits when he finds that plagiarism is so prevalent that it has become synonymous with research.[458] He finds the last straw on proverbial back of the camel in the form of reports that even authorship of scientific papers is for sale.[459] He does not know as to what should he do. He needs answers.

The organisation of science and technology in post-independence India is discussed by Rahman et. al.[460] Organisation chart of various scientific departments and their distribution in various ministries is shown in a chart.[461] These include the autonomous bodies, Central Government Ministries and departments, State Government departments, Industry and private institutes.[462] The channels of communications between the Government and the scientific establishments are demarcated. Policy decisions are conveyed through either the cabinet secretariat or the ministerial functionaries. The scientific departments come under individual ministries headed by a minister and has a secretary level officer to oversee the administration. The department of Atomic Energy has traditionally been directly under the Prime Minister with an independent secretariat and a commission. In many ministries the secretary is a career service officer from the Indian administrative Service. The organisational structure of Indian Council of Agriculture Research (ICAR)[463], Indian Council of Medical Research (ICMR)[464], Council of Scientific and Industrial Research (CSIR)[465], Defence Research and Development Organisation (DRDO)[466], and Department of Atomic Energy (DAE)[467] are presented by Rehman. These are headed by scientist/administrators. The Indian Space Research Organisation (ISRO)[468] was started as a part of DAE but charted an independent course later. Each node of the hierarchy is manned by a scientist/administrator. DRDO has an organisational structure that is a bit different. The institutes under DRDO have an army officer and a Chief Scientist as head, one looking after the administration and other responsible for the scientific activity.[469] The organisational structure of all these departments is hierarchical and acts as conduit for the policy decisions of the government. At the lowest rung of this are the individual institutes which are tasked with fulfilling the scientific objectives set forth by the policy makers. The administrative structure within each institute, except for those under DRDO, by and large are identical. All these are hierarchical in nature.

In order to understand the ecosystem of a typical hierarchical scientific institution we will take as example, the case of BARC for reason none other than the author's familiarity with it. Data about its employees in different categories, organisational structure, delegated powers and the decision-making process is available from open sources. The Atomic Energy Commission has several research institutes, public sector corporations, and aided institutes.[470] The research institutes like BARC have an administrative structure that subdivides the manpower and arranges them in Groups, Divisions, and Sections. There are currently 19 Group Directors, 17 Associate Directors and 93 division heads.[471] The group of scientists gracing these positions control the activities of about 4665 gazetted officers (mostly scientists) and 9979 non-gazetted (Technical & administrative) supporting staff. [472] A Group may have several divisions and each division may have on an average 5 to 10 sections. A section may have many working groups with on average 4 to 5 scientists. This lower most unit, the working group and the scientists comprising it has the highest responsibility of fulfilling the scientific and technical objectives set forth by the chain of command starting from the Minister-in charge.

The delegation of powers to scientists holding different positions in the hierarchy shows successively diminishing administrative and financial powers with the head of a section at the lowest rug. [473] The financial power at his disposal is of only Rs. 5000, whereas the vast majority of scientists do not have even this. The position of section head, a quasi-administrative position, may be offered after about 5 to 10 years of work as a scientist. The person holding this position is still scientifically active and his administrative load is minimum. However, as he moves up the administrative ladder his active involvement with science reduces.

The decision-making process, as recorded on the BARC website, is as follows:[474]

"The Centre is organised into various Groups and Divisions/Sections and co-ordination of its activities is achieved through the Group Boards, Trombay Scientific Committee and the Trombay Council. Director is the Head of the Unit. He is also declared as Head of the Department and has been delegated financial, administrative and other powers by the Department of Atomic Energy for carrying on with the activities of the Centre. Other functionaries like Group Directors, Associate Directors and Heads of Divisions/Sections have also been delegated powers required for discharging their duties/functions. While the Group Directors, Associate Directors and Heads of Divisions are competent to take decisions in their delegated areas of responsibility, inter-divisional co-ordination is achieved through the Group Boards and Trombay Scientific Committee. The Trombay Council is the policy making body for the Centre. The deliberations of Group Boards, TSC and TC are not open to the public and minutes of such meetings are also not accessible for public".

It is clear from the numbers that that about 2-3% of the science administrators who participate in the decision-making process control the activities and fate of the rest of the working scientists. Even if one takes into account the number employed in the purchase[475] and administrative support departments the above figure will marginally change. Thus, it is clear that the large population of active scientists have no powers, financial or otherwise to cater to the needs of scientific work and have no voice in the decision-making process. BARC website states that the deliberations of the policy making bodies like Trombay Council (TC) and Trombay Scientific Committee (TSC) are not open to the general public. The scientific community of BARC comes to know of the decisions by trickle down process if the concerned chain of command deems fit. On the ground, a working scientist's voice even on scientific matters affecting him is not heard and there is no scope of disagreement with higherups. Even if one takes the liberty of expressing his opinion frankly, the Damocles sword of annual confidential report (ACR) comes heavily on the unfortunate subject.

The scientist has to work in an ecosystem which is full of committees. It has a labyrinth of committees, procedures and rules through which a working scientist has to meander his proposals for procurement of equipment and consumables before he can settle down to do some worthwhile research. The procurement of equipment is governed by the financial rules[476] and procurement rules[477] set up by the government. A major portion of scientist's time thus is spent in chasing the purchase files through the committees and the purchase department, getting his designs fabricated from the workshops and cajoling and persuading the technical staff for completing the jobs assigned to them. He gets squeezed from both sides; for fulfilling the objectives from top and from the bottom by indifferent attitude of the technical staff who often fails to relate himself to scientist's objectives. It is also a fact that ratio of the number of technical staffs to the scientists is not optimal. It is often left to the scientist himself to also do the menial job that comes in the realm of the duties of the supporting staff. As a result a scientist finds less and less time for keeping himself abreast of the latest trends in any fast-moving arena of research. The delay in getting his equipment and experimental system leaves him running after a mirage and the objectives and goals of the department recede and fade into oblivion.

It should also be recognized that a scientist in India lives in a feudal social milieu. Feudalism in science is rampant in almost all the scientific institutions in India. BARC is no exception. A junior scientist has to share the toils of his work in the form of giving authorship to administrative heads lest his career growth prospects are adversely affected. Ignoring the hierarchy might also lead to withdrawals of funds, equipment, facility and even laboratory space. There are subtle ways to enforce compliance. Invoking confidentiality clause to deny permission to publish is one method. Some heads use administrative procedure of getting clearance for publication even if the work is not of strategic importance or of confidential nature. In science, where timing of the publication is important, a scientist who has waited enough for getting his equipment, is eager to publish his work as soon as possible. If the head of department's name is not in the authors list of the paper submitted, he may hold on to the draft for months on end. If suitable corrective action is not taken by the scientist, the permission for publishing may also be withheld. If the head is one of the authors, the draft will promptly be returned with minor corrections or suggestions. If the scientist sends the draft paper with his superior's name as the first author in order to speed up the process, the draft will be promptly returned with approval, with superior benevolently shifting his name to the last position. A cursory reading of the last para, of any paper, makes an appropriate comment on the psyche of the scientist. The supine manner in which the words of gratitude for the head of the department that flow with such phrases: 'for allowing the authors to work', 'for constant encouragement', 'for providing facilities' etc. is akin to prostration. Added to this is professional jealousy that controls the behaviour of fellow travellers and the controlling authorities. In a typical example, a scientist publishes, within a short span of time, three papers in a prestigious journal, only to find the laboratory space and equipment taken away. Another one works hard on a project, nearing completion, only to find that he is asked to hand it over to someone else. The scientist is left to himself, no work assigned, no questions asked. The fellow scientists ostracize him and keep a distance for fear of repercussions. In this situation some would leave in frustration and migrate, an avoidable loss to Indian science. Exemplary examples like these create an atmosphere of fear in which consciously performing below one's capabilities becomes a norm, lest it may invite some attention and retribution. Mediocrity is a boon that has to be consciously cultivated to survive and flourish in the hierarchical knowledge system of government science. This is the ecosystem in which excellence is decried and tacitly supported by the silence of the community.

The scientist, once thermalized to the prevalent ecosystem, as he moves up the administrative ladder, finds himself saddled with host of committees, both within and outside the organisation, leaving very little time for his true vocation. Lure of social status within the scientific community and a feeling of power to control the fate of fellow scientists drives him to gladly accept to be part of a host of inconsequential committees. As time passes his managerial responsibilities increase with position, but progression of his carrier requires proof of his scientific contribution. No wonder that he has to resort to such coercive actions to get his name as an author in the work of junior colleagues. Generally initial ten years of one's career by and large, are enough to define one's field of work. As the scientist moves up the administrative chain more groups, working on areas of research other than his own, come under his umbrella. List of scientist's papers in fields diverse from his own field of expertise keep increasing. Such inflated list then becomes a basis for state recognition, appointments and awards. Very few realise that such a behaviour comes under unethical authorship and is against the grain of science.

Rethinaraj and Chakravarty[478] has brought out such unethical authorship in DAE institutions and highlight the case of a former IGCAR director. Rethinaraj and Chakravarty write;

> *"Raj has a prolific publication record of 714 Scopus tracked peer reviewed works but more than 1300 publications according to his recent curriculum vitae. His productivity level is exceptionally unusual when compared to any other leaders of DAE establishments. A summary of our investigation using data from Scopus is depicted in the figure below. It clearly shows that Raj's publication record is unique. In 2011, the last year of his tenure as Director of IGCAR, Raj's research productivity peaked at 77 publications (one every 4.7 days). Raj's tenure as IGCAR Director (2004-2011) were his most productive years during which he authored 388 of his 714 Scopus tracked articles, which represent 54% of his lifetime output".*

Authors also show that during the years when Baldev Raj was in charge of Materials and Metallurgy Group (MMG) his average annual productivity was 10 however when Chemicals and Reprocessing Group (CRG) was added to his charge the same parameter rose to 22.[479]

Baldev Raj, the Ex-IGCAR director defended his record and said that it is not exceptional.[480]

> *"Publication authorship is ethical commitment between authors and to my mind, no one has a right to comment on this. Only authors are responsible for ethics. My publication record is eminent but not unachievable by scientist."*

The ethical commitment that Baldev Raj talks about is between scientists of equal status and not between the boss and a subordinate in a hierarchical administrative system. A subordinate works under constraints of the hierarchy and the weight of the annual confidential report that his superior decides. Hence it is not a marriage of equals. What happens in practice is nothing but the 'Director effect'. This effect, in varying degrees, exist at all levels of the hierarchical science administration. As a scientist moves up the ladder, he gets more disciplines under his wings and the fruits of the research by his charge are duly shared. There is no mechanism to check if the person concerned is expert in all the fields as shown by the list of publication in his CV. The entire chain of command simply ignores it as it has become a norm.

The result of the exposure by Rethinaraj and Chakravarty led to disciplinary action against them and the tragedy of Indian science is that the larger community of researchers remained passive observers.

Unethical authorship is a demeaning behaviour and acts as discouragement to good work and cynicisms in the minds of a new entrant to scientific research. He sees before him his whole career subservient to the bosses until he himself becomes one. He grudgingly parts with the credit of his labour instead of protesting for fear of his career prospects and gladly accepts tributes later in his carrier.

The fist-fight between the two BARC scientist,[481] that we referred to earlier, had its origin in charges of plagiarism that led to retraction of papers by the concerned journal.[482] The total number of papers retracted is said to be about seven.[483] The incident and its aftermath poses some questions that needs to be answered. What was the response of the administration? The complainant was harassed by removing his laboratory facilities.[484] What action was taken against the scientist involved in the plagiarism, a charge established to be true by the withdrawal of the papers by the journal? It is surprising that it did not invite any scrutiny of the particular scientist's entire record of publications. Instead he was promoted to a higher position.[485] This is in sharp contrast to strict action taken against a senior scientist of the Institute of Microbial Technology (IMTECH) who was removed from the post.[486]

The incident raises few questions. Why was the defaulting scientist promoted after the retraction of his papers came into light? Logical action would have been removal of all those retracted publications from the service record of the concerned officer and all benefit given to him on the basis of his past performance also retracted. The department would have set a proper example and gained prestige in the community of scientists. Was any action taken against the co-authors of the senior scientist concerned? In all probability, they too would have been promoted. As the lead author was promoted, there is no reason to believe otherwise. The co-authors became an unquestioning accomplice in this unethical practice. As scientists, the least they could have done was to dissociate themselves from sharing the credit. In all probability this was the labour of their efforts, attachment to which might have clouded their thinking. It may also be that they themselves were unaware of the ethics of scientific research. It would have been a difficult decision but certainly would have raised their stock in the eyes of fellow scientists even though it would have been termed foolish by the diaspora. Did the administration under the guise of service rules punish the whistle-blower or was he awarded for his noble deed of exposing a fraud? We do not know.

Contrast this with the censure that Watson, of double helix fame, had to face for his unscientific remark about race and intelligence.[487] The institution acted without regard to the scientist's contribution to science or his international fame and stripped him from final honorary roles of the institute.

In the light of the fact that incidences of plagiarism are on the rise it is necessary to sensitise the junior scientists to the negative impact it can have on their careers. The departments should come out with guidelines for participation as author for all scientists holding administrative positions. The scientific officers' associations have a role to play, in the interest of the community, to be watchful of the wrongdoings.

Associations in institutions suffer from a general problem of credibility as a perception goes around the community that only the disgruntled join it to hide their unwanted activities, further their own ends or to curry favour from the administration. It is also true that the associations can become the last refuge of scoundrels as rouge employees might find immunity from the administrative actions. The scientific community has a responsibility of sending persons of impeachable credentials to the associations. The officers' associations have a responsibility and a major role to play in the interest of the Indian Science. It should in fact be realised by the administration, the unions and the associations that the stakes for ethical working are rather high for Indian science to gain a place of honour in the society.

The ostrich response of the Indian Scientific Societies towards plagiarism and unethical authorships by their members is worrying. These organisations are supposed to be conscience keepers of the community. Did any scientific society revoke the membership of any deviant vice-chancellor or scientist who was found to have indulged in plagiarism?

Another factor that affects the ecosystem is a conflict between pure and applied research, a factor, that was recognised to be present in defence organisation quite early.[488]

"The compartmentalisation that existed between academicians and the technical personnel engaged in defence work, the perception that scientific work in defence is only of applied nature, the sharp distinction that was sought to be made in academic circles between pure and applied research with disadvantage to the latter ("a distinction nearly as sharp as the distinction between a gentleman and a liar at large")".

S. S. Bhatnagar, H. J. Bhabha and K. S. Krishnan, with D. S. Kothari as chairman, were the members of the advisory board that was to formulate the science policy for defence. However, there is no evidence to show that any concrete steps were taken in any of the multitude of research institutions that were established under the tutelage of these eminent science administrators to address this problem. The dichotomy between the pure and the applied research remains in the ecosystem to this date. It is used often by various committees dealing with the assessment of work.

How this distinction came into existence and then crept into the administration of government science, post-independence is a matter worth paying attention to. Is it the outcome of emphasis by metropolitan scientists and administrators of the imperial era placing more emphasis on the commercial aspect of the scientist's foray into the colonies? It is known that the scientists straying from commercial objectives were severely reprimanded. Isn't the fact that almost all the British Noble laurates were metropolitan scientists, support the view that applied research was for the colonies?

The scientific structure of DRDO and CSIR adopted by post-independence India was in continuum with the past structure. The centralised planning and control of science led to centre-periphery relation vis a vis the funding agencies at the central government and research labs scattered throughout the country. In the context of DAE, the fundamental research at TIFR and the applied research in other laboratories of the department and preferential treatment to former led to a chasm between the two streams within the department. The stepmotherly treatment given to BARC as compared to TIFR, by the founding father might have led to this divide. The conflict between the scientist engaged in the pure research and the applied research erupted quite early over the sharing the particle accelerator, imported at TIFR in early years.[489] The difference in philosophy for establishing TIFR and the AEET, Trombay (BARC) might also have contributed to this dissonance.

Another aspect that probably got insignificant attention of the academician and historians is how the appointment of famous scientists in the key positions might have affected the post-independence ecosystem of science. Without negating the contribution and untiring efforts put up by the four key scientists involved in planning and establishing the physical infrastructure, one should look at the deficiencies as well. Did the Prime Minister of the time pick up right persons for the job? Did he have enough foresight of the manner in which science was to progress? Which model for scientific growth was the country to follow? Was it assumed that given the right infrastructure the chosen scientists will themselves evolve the right environment? S. S. Bhatnagar was a chemist well versed in industrial chemistry. When he took charge of CSIR (BISR) he brought his research group with him and tried to establish a laboratory but found little time for research. D. S. Kothari was a theoretical astro-physicist and Dean of science at Delhi University when he was picked up to head Defence Research and Development Organisation, an organisation whose work was far removed from his expertise.

"Dr Kothari worked in the area of quantum statistical mechanics and its applications to degenerate stars and planets."[490]

How did this affect the growth of defence research? Defence research to a large extent and civilian scientific research to a lesser extent was influenced greatly by the advice given to the prime minister by P. M. S. Blackett. Blackett was a frequent visitor to civil and defence establishment and senior generals and scientists interacted with him. Blackett was active in Indian science until his death in 1974. He was asked to study and report on the functioning of CSIR and his views were forthright but almost ignored by the science administrators who resorted to a simple action of removing the in-charge of the institution rather than addressing the core problems of the stifling atmosphere of rules and regulation and absence of free expression. Another point of consternation, that often arises in scientific institutions between, the scientist recruited directly and those coming through departmental training, is the bias in favour of inhouse trained officers even though they may be assigned to work on the same problem. The carrier progression for the two streams are different. This breeds resentment and non-cooperation between the two. Yet another differentiation is between physical sciences and the engineering disciplines with balance tilted in favour of the engineering sciences even if both work on the same project. The scientists of the two streams have different growth progression with engineers surging ahead at a faster rate. What it does to the scientific atmosphere? There is lack of cooperation and exchange of ideas and encourages protecting of individual spaces.

The "Blue Eyed Boy" syndrome, that at times cloud the judgement of the senior scientists holding high positions in favour of an individual at the cost of science, creates a depressive atmosphere for a creative individual. Science flourishes in free and open atmosphere. If a scientist cannot voice his opinion even on scientific matters, lest it may attract retribution, calls for a serious introspection on the part of science planners, scientists and the administrators.

The desire for an administrative position and the lure of power associated with it kills the creative urge of a scientist. The scientific output of many famous scientists declined or even stopped due to the demands on their time for administrative work. Bhabha's is an excellent example. He was an internationally renowned scientist when he founded TIFR and in about ten years' time his scientific output declined to nil. His last scientific paper was published in 1956. He was only forty-seven years of age. Biographers make a virtue of it, by terming it as a sacrifice for larger interest of the country. The fact remains that, this was a huge loss to science in general and Indian science in particular. This was a loss that a nascent country could only ill afford.

It is always tempting to ask a 'What If?' question, even if a futile one. What if Bhabha would have taken up an academic position in any university instead of ambitiously creating a humongous organisation that has failed to replicate even a single scientist of his own calibre even after more than half a century of its existence? Who knows he would have nurtured at least one if not more world class scientist. If Sir J. C. Bose could mentor M. N. Saha and S. N. Bose in the university with limited resources, why not Bhabha with huge political backing and resources at his command?

Let us now recall the contributions of T. R. Seshadri, N. R. Dhar and C. V. Raman about whom we mentioned in passing [Chapter2]. Seshadri declined chairmanship of the University Grants Commission and worked in various universities, setting up new departments and mentoring about 160 students for their Doctorate degree in his career.[491] N. R. Dhar, student of Sir J. C. Bose and P. C. Ray, is known to have contributed significantly to soil science. He was an university academic and guided about 150 students.[492] G. N. Ramachandran[493], a student of Sir Raman had a distinguished scientific career and is known for his work on X-ray crystallography, Ramachandran plot[494] for peptide structure and triple helix for collagen. Considering the contributions of these university scientists the 'What if' question has a relevance. If only the government had invested the same amount in the universities instead of the big organisations the results could have been different.

The stark fact that a young scientist faces is that the founding fathers of government science, in an attempt to free themselves from the bureaucracy, forgot to free the working scientists from the scientific bureaucracy that they created. The power that was concentrated in the top few individuals rarely percolated to the working scientists. These powers, as we have seen, were often marginally delegated lower down thus handicapping the working scientists.

The government's attitude toward giving importance to its scientific cadre can be inferred from a simple comparison of the officers from two cadres, namely IAS and the scientific. A graduate entering the Indian administrative academy after passing out is placed as District Magistrate with substantial financial and administrative power. Contrary to this a scientist, with similar or even higher educational qualifications, does not have any power to fulfil the objectives assigned to him. At places, scientist does not have freedom to withdraw even elementary necessities like pencil and paper from stores without approval from his superiors. This inevitably leads to frustration. Young scientist even from a prestigious institute like BARC are known to   prepare for IAS and if selected leave scientific research.

In this eco-system, at the bottom of this administrative ladder sits the young scientist, freshly minted from the almost defunct university system. He comes from a society where faith excludes any questioning, where respect for God, teachers and elders, reinforces their obeisance, where appeasing the powerful through offerings is a norm. The education has not prepared this young scientist to face the challenges of honest scientific enquiry. He is partially educated as he was at no stage, taught about the history of science or it's values, ethos and the principles, except being exposed to eminent scientists of the yore and their discoveries, through textbook hagiographies. The essential elements of scientific methodology are not taught by his teachers at any level of his education. Nor are the teachers enlightened enough to gauge its significance. Even during his training, as prelude to formal employment in any scientific organisation, he is not exposed to the history of science, methodology and ethics. He has not heard who Francis Bacon was and what was his contribution. How western science progressed during the European scientific revolutions, he does not know. He learns the methods of enquiry while on the job and acquires the attitudes and practices prevalent in his immediate environment and develops a value system from the peers and from the ecosystem. Very often he comes from a middle-class family background that has traditionally shunned manual labour. The rudimentary arts and crafts classes in his school days have not imbued in him an appreciation of the menial work. Since the society from which he gathers his social values abhor the menial work, he carries with him the same attitudes. The euphoria of being a class I gazetted officer of the Government of India precludes him from undertaking any manual work himself and has to rely on the whims of his assistants, technicians and helpers for day to day work. This young scientist is the foot-soldier for the government enterprise that is glorified in the name of science but is hardly different than other arms of the government. He starts with being marginalised in the decision-making process and works his way up the administrative ladder until he reaches the position where he is acclimatised to the manner of governance about which he himself was critical in his days as young scientist. His efforts, then, are directed towards protecting his turf and working towards his personal gains in the system. This is what stares at the young aspiring scientist's face.

We now come to the question - Are we heading somewhere in science? what is its future in India?

Olson cites two conflicting lines of thinking on the meaning and characterisation of science-[495]

" .....*a group of British Marxist scientists and interpreters of science, including J. D. Bernal, Benjamin Farrington and Joseph Needham, who focused on <u>social and political contexts for the development and uses of science</u> on the one hand and a group of scientists and students of science, including Michael Polanyi, Robert Merton, and Vannevar Bush, who sought to focus on the <u>freedom of good science from politics and social needs</u> on the other".* (underline mine)

The importance of progress in science and technology for country's development was recognised early by Pt. Nehru who initiated the consultation with the scientific class even before independence. He initiated science planning early and was personally involved in the deliberations. He was probably influenced by Bernal's thinking. Did the scientists entrusted with the task had faith in his dictum or followed the other view? Was the society ready to come out of age-old dogmas and beliefs to imbue the scientific temper?

The freedom movement in India resulted in adult franchise and a democratic polity replaced the Imperial government. Though the constitution showcased secularism as the state policy, the effect of religion and the obscure practices continue to dominate the social discourse. Appreciation of eastern religions, philosophy and culture in the guise of Orientalism[496] gave the Indians a sense of superiority of the past over the materialism of the west. Spirituality again started asserting and the scientific culture which was beginning to seep in the Indian society before independence was pushed to the background. Mechanisations of electoral compulsions resulted in the pandering of vote banks and patronising of the religious gurus by the political parties, giving them an added advantage. Politicians routinely paid obeisance to religious heads, offering poojas at temples and chadors at the dargahs reinforcing the belief of uneducated masses in their religions. Spectre of renowned men of science sitting at the feet of godmen added to double whammy and the disillusionment of public with science became complete. Political parties started relying more on saints and sadhus. When sadhus and saints, who had renounced worldly gains in favour of spiritual quest, realised that the politicians are enjoying the fruits of power utilising the devotion of their followers, they themselves entered politics and became legislators, member of parliaments and ministers. The society, over the years saw an increase in religious discourse, yatras, pravachans (sermons), Satsang, urus and host of religious functions, organised with increasing grandeur. Splendour of rich lifestyles of the religious preachers further marginalised science in the minds of the population. After independence, many preachers forayed into the western societies and gained wealth and prestige abroad and at home. Orientalism did spread the mystique of the eastern civilisation in western world. This enforced belief that if west is falling head over toe at the feet of neo gurus, who need science for uplift of the society? If wealth is no longer a bane and is a means to attain salvation, why separate it from religion. Renouncing worldly gains is no longer the essential perquisite for attaining moksha. When uneducated babas (hermits) can use religion for setting up business worth thousands of crores of rupees, make forays into Bollywood, science has lost relevance for the society. When governments accord cabinet minister's designation with all the perks and privileges to sadhus, give doles to mendicants, patronise dharam-sansads (religious congregation), and prepare pujaris and Jyotishies (astrologers) in the educational institutions, who needs scientific temper?

Let us now address as to how science delivers. In early twentieth century two linear models of the innovation process namely, Science-push model and Demand–push model, were in vogue[497]. In Science-push model developments in basic sciences push R&D to develop products which were then pushed to production and to the market. In demand-pull model the market is the source of ideas for directing R &D and then follow the rest of the steps to market.[498] Later several modifications to this linear model were proposed to include feedback from other factors like the need, economics etc in the system. Godin and Lane write;[499]

"In the past thirty years or so demand was diverted to mean economic demand and was peeled off from need. Yet, supply (scientific discoveries) would play a different role in theories and policies than it currently plays if models placed the emphasis on needs and the beneficiaries of innovation. Rather than the dichotomy of either universities or firms being the drivers of innovation systems around which other participants play the role of "context", the emphasis would be on

1. consumers, citizens and their community associations,

2. public managers and programs,

3. governments, public organizations and policies.

This is what J. D. Bernal suggested in 1939, although from a normative rather than an analytical point of view, "If we take human life and its development as the center of our study, the activities of science assume a different aspect" (Bernal 1939: 345)".

In the western countries the onus of scientific research, largely, is on the universities and the discoveries and inventions flow to the market through private enterprise. On what model, in India, the centralised research institutes were established? CSIR was to cater to the various industrial sectors. Did its institutions have the end product in view and was its demand assessed? Or the science in these labs was to lead to products for commercial exploitation as in science driven model? Over emphasis on starting R & D institutes with the fond hope that some products will emerge led to a situation of a horse running after the cart. The pace at which International science was then moving towards meeting the challenges of increasingly refined technological products ensured that the Indian science with rudimentary technological infrastructure, moved to obsolesce even before it started walking.

The research at DAE had its nuclear end product in sight and the research being conducted in these organisations largely was of applied nature catering to the commerce driven needs. Geo-political importance of possessing a nuclear weapon must have been on the back of the political minds. The research in areas that were not related to the objectives in which research was started was termed as pure, creating internal divisions and the rifts. Between the fundamental research of cosmic ray variety and applied research of the atomic energy type, an oceanic chasm of the absence of basic science research developed.

Research in basic science at the universities, that could have been the harbinger of innovative technology, was missing. What remained intact of the last stage of linear innovation chain was the Indian market. Counting on the technological benefits, arising out of science elsewhere, the country could easily use globalisation to its benefit. With the technologically advanced industrial countries eagerly willing to exploit huge population and supply products to a fast-growing Indian market, why does India need scientific research at all?

With the opening of global trade, what was needed was infusion of marketing education, promoted by western countries. This led to mushrooming of management institutes in every nook and corner of the country as the multinationals needed cheap local managers to manage the flow of goods and services. These management institutes taught the curriculum written by western counties from the books written by the westerners. Also, if products of western science have to find a market the purchasing power of the society has to be strengthened. This was achieved through business process outsourcing, drudgeries of the western societies, to BPO's The under educated English-speaking students got an opportunity to work at emoluments that they could not have dreamed with their qualifications. The educated engineers and graduates of all varieties found employment as sales and marketing employees and managers. Electronics and computer engineers became software developers for western companies and the governments. No electronics hardware industry could develop. Government tried to cope up by setting up government institutes for chip design of higher and higher component densities but in no time went into obsolescence as these could not match the pace of western technological progress. Infusion of higher than the average emoluments to service jobs raised the local demand of the technologically advanced products of western science. The basic science and the innovation cycles could not be established and opening of global trade subsumed everything under the deluge of west originated products.

The dismal education system and lack of opportunities within the country led to migration of talented Indian students for study and research in basic sciences in the western universities. The products of their labour are now channelled, via the Chinese manufacturing, to the emerging markets like India. The cycle of linear model of the use of science is completed, with Indian science missing from it. Science found an ornamental value with government institutes becoming temples of modern India. In this scenario why crib for the absence of scientific temper in public discourse when the scientists themselves do not bid for it?[500]

We may as well pray for the departed.
*

11

Is there a way?

As we have seen in previous chapters, the odds are stacked against the growth of modern Indian science. There does not seem to be clear way out. However, inaction is not an option and some of the concerned individuals keep suggesting ways to come out of the morasses.

Joseph and Robinson in a letter in *Nature* has forcefully advocated for removing government shackles from scientific institutions.[501] This is not the only thing. Lack of accountability, ethical values and morals amongst the scientists is a serious matter that needs to be addressed apart from the structural changes in the governance. The present hierarchical system needs a reform is an undisputed fact as it has so far not served the purpose. Let us analyze the situation and start by recognizing that there are two streams of scientists, one needed for the administration and second for active research. It will be advantageous to create a scientific administrative service, like IAS. It will be prudent to offer a carrier path diversion to mid-level scientists to cease research and occupy managerial positions. This will free the person holding the post from the need to exploit the subordinate scientists for his own career growth. He should be offered time bound career advancement, evaluation of which will be on his ability to manage the infrastructure and facilitate the scientists in their research. His performance will be linked to the citation index of the publications from the institute, no of patents awarded to the scientists and timely completion of the projects.

The scientist opting to remain active in research should be given all the financial and procurement powers which he can exercise independently of the administrator. Delegation of financial powers, coupled to the grade, should trickle down to a scientist at the lowest step. The allocation of funds also should be suitably done amongst the scientists at various levels. The work of scientists should be evaluated by the citation index of their work or projects successively completed without cost and time overrun. Claims of projects completed should be evaluated by external peers.

The research outside the central institutions is now limited to the universities. The research in Indian industry is minimal and is used for tax exemptions. The central and state governments fund research projects in the universities. These agencies face a problem while assessing the suitability of the research group for funding. A recent study brings out the unsuitability of existing criteria based on the citation index of the authors work and that of the publishing journal for deciding the suitability due to nonuniformity in citation indices.[502] In order to overcome this nonuniformity it will be advantageous if all authors are asked to publish in Indian journals. This will remove at least one major factor, variation of citation index of foreign journals. It will also be advantageous to the journals if the funding agencies use citation index of these papers as the basis for future funding.

It is well-known that Indian authors generally prefer foreign journals for their first-rate papers and Indian journal for the rest. In most of the institutions the assessment reports want separate list of the publications in foreign, local journals and symposium proceedings. Over emphasis on publications in foreign journals amounts to self-degradation. The result is that the Indian journals, baring few, fare very low on the world citation index. Often better international exposure is cited, by researchers, as a reason for the preference. This apprehension, in the minds of the scientific worker, would have been valid before the advent of internet. With a host of online platforms now available this reasoning loses sheen. The Indian periodicals should be made freely available online. If a scientist still chooses to publish in foreign journal he should be free to do so but his funding from the government should depend only on the citation index of his published work in Indian periodicals. Advantage of compulsorily publishing in Indian journals, over time, will not only boost the citation index of Indian journals but also provide the funding agencies some rational basis for awarding grants. The Indian journals also need to change their working and outlook. The publications should be out on schedule. For better content at least one international pear reviewer should be made mandatory. That will further enhance the quality of the publications.

Since the universities form the backbone of western science and technology it is important to reinvigorate the universities. The universities should be made autonomous with no government control and political interference in all matters including academic appointments, curriculum and research. Deepak

Kumar has proposed linking of research at universities and the government laboratories.[503] The attachment of central laboratories to the nearest universities can be a positive first step that can be mutually beneficial. This is necessary for infusion of fresh minds from the universities into the central institutes, crossflow of ideas and utilization of expertise available in the central institutes.

In recent times, the government started several institutes of education and research (IISER) on the lines of the Indian Institutes of technology, devoted to physical sciences to boost declining numbers of students opting for science streams.[504] It is too early to assess the true impact of such a move on science education and research. These institutions, like IIT's are not universities and lack horizontal knowledge dissemination so central to universities of the west and other advanced countries. Also, the fact that many of these new institutions are staffed in their infancy by scientists, either retired or on deputation from the existing government institutions, carries a real danger of the same old culture and ethos permeate in the new ones. One of the IISER prominently claims that most of the graduating students find opportunities to study in universities abroad.[505] If this is the Unique Selling Proposition (USP) of the new institutes, the purpose of invigorating science education and research is lost.

The existing research institutions, run as typical government organisation, are unwieldy in terms of manpower and resources, leading to inefficient functioning. The culture of indifference and casualness does not bode well for any change. It will not be possible to drastically change the present institutional structure in the democratic setup. If persuasion does not yield results the existing institutes should be allowed to function with minimum support.

How did China despite losing decades in internal strife advance so fast in science and technology? China began by sponsoring education of a large number of students in western countries and then created dedicated research institutions for Chinese scientists to return and work. A large number of qualified overseas scientists returned to the new institutes. All these institutes functioned under the Chinese Academy of Sciences. The result of such an organized effort was 17% increase in world share of publications from China in about two years.

In India, a new institutional structure and mindset therefore is needed. The new institutions should be small and autonomous institutions, unshackled from government rules and regulations and registered as societies with governing bodies consisting of a group of eminent scientists in the field, relevant to the institute. These institutes should be attached to the universities and should participate in its teaching and research programmes and take students on their research problems. The staff at all levels should be on fixed term contract, renewable after performance appraisal every five years. All powers should be vested in the working scientists who will also have to seek funds from the industries and the government agencies by writing proposals. The resources allotted to the scientists should be weighed against the products and services delivered by them in addition to the citation index of his publications. In the present system a scientist's time is taken up in submitting periodical reports and answering various queries. This should be minimized. The benchmark should only be the quality of work. However implementation of these suggestions not only depends on the sincerity of the government and the political parties in weaning away the dogmas from obscurantist dogmas and religious beliefs but also on the scientists themselves to be true to their profession.

**

Notes

Note-1
{ref. 99}

".... In October, he wrote both to Russell and to the physicist Robert A. Millikan asking for assistance "from some charitable American Institute" to purchase a good ultraviolet quartz spectrograph, diffraction grating, transformer, and other spectroscopic accessories. ......Russel was sympathetic but differed to Millikan, who was an adviser for physics to the Rockefeller philanthropy."

".... But Millikan had gained a low opinion of Saha's potential from C. V. Raman, the Indian physicist. Saha had escaped in Calcutta, who was then visiting Caltech for the autumn term. Although Millikan reported that Raman had told him that Saha had done "some excellent theoretical work", Raman also claimed that Saha was "in no way an experimentalist or an organizer, and funds spent in the way Saha had requested would not be likely, he feared, to represent the wisest expenditure which could be found" Raman, according to Millikan, stated that "if Saha had inspired confidence in India in his ability to get results through organization and direction of research he was confident that it would not have been necessary for him to apply in this country".

**

Note-2
{ref. 108}

"After the death of Mahendra Lal Sircar's son Amrit Lal Sircar in 1919, Raman was made the secretary of the Science Association and its chief executive, Over the years his control extended to all spheres of the association's activities. He had co-opted in its Committee of management a large number of members of his choice. Both wife Loksundari and brother C Subrahmanyam Ayyar were

on the committee.   .... He was at the pinnacle of his achievements, his position was unassailable, and enjoyed immense freedom to do things his way....".

".... however, there were rumblings of dissent. A feeling was gaining ground among the younger scientific community that Raman was building a South-Indian coterie around himself, ignoring the claims of the native Bengalis. ...... ".

**

Note -3
{ref 109}

Mallik D C'' On 20 May 1933, the management committee of the Science Association met and formalized the institution of the MLS Professorship. In the absence of President Sir R N Mukherjee, Dr Sudhanshu Kumar Banerji, chaired the meeting. .... The committee noted that it had received Sir R N Mukherjee's letter of resignation from the presidentship of the association. .... In the same meeting, C Subrahmanyam Ayyar and K Rama Pai proposed that Raman should take over as president of the association in the vacancy created by Sir R N's resignation. As soon as this was accepted, Raman proposed Krishnan as the new secretary of the association. This motion too was carried through, without a murmur. Thus, in two quick moves Raman's control over the affairs of the association. In absentia, was secured. However, in the same meeting when the names of Syama Prasad Mookerjee, Rama Prasad Mookerjee and Jnanendra Nath Mukherjee were proposed for election as members of the association, the proposal was defeated".

**

Note -4
{ref. 110}

"None present in the management committee meeting that evening could have foreseen the events that followed in quick succession ones the news of the June 19 meeting spread. Syama Prasad, with his acumen, immediately saw the chance of setting matters right for himself and the science association. He marshalled all the support he could, and with the help of Meghnad Saha, then *on vacation in Calcutta from Allahabad, inducted no less than 68 life-members in a span of few days by extracting donations of Rs 250 from each of them. .... Suddenly, a host of Bengali gentlemen, who had never before shown any interest in the affairs of the Science Association and many of them perhaps without any specific interest in science, found themselves drawn into the centre-stage of a drama that was unfolding, ...The storm gathered and when the time for the meeting arrived, Raman was stunned to see so many faces in the hall, faces he had never seen and names he had never heard in his 27years of association with the institution. He hesitated to enter the meeting hall and even tried to prevent Krishnan, the secretary of the association, from doing so. But Syama Prasad was unstoppable and after some acrimonious exchange of words and after Raman was shown the receipts by the trusties, succeeded in forcing Raman to attend the meeting. The game was over........ ".*

**

Note -5
{ref. 113}

"Towards the end of 1932 there were letters to the editor in the local newspaper against Raman's management of the IACS and the Palit Professorship. The accusations were that he had only south Indians around himself as scholars, and that physics was given too much prominence, to the exclusion of other sciences. The main grouse was that Bengalis were being side lined in their own province".

**

Note -6
{ref. 126}

"Raman requested A S Ganesan, a professor of physics at the Science Association, to compile a bibliography of papers on the Raman effect complete up to July 1929.14 Sircar relates an interesting story about the manuscript of this bibliography, which was printed and circulated separately prior to its publication in a regular issue of the Indian Journal of Physics".
"After the bibliography had been written Professor Raman wanted to publish it within a few days. So he asked some of us to communicate each a paper to

Indian Journal of Physics as early as possible so that the bibliography could be published in an issue of the journal......".

"Professor Raman asked Bhagavantam, myself and a third author to accompany him to the old Senate hall at the north west corner of which the office of the Superintendent of the Calcutta University Press was situated. We went there with him and he handed over the whole matter to Shri Autosh Ghatak who was the superintendent at that time and requested him to set up the matter immediately and give us galley proofs, so that they could be corrected by us sitting in his office.....within two hours we got the galley proofs for the whole matter. .... After being corrected, the galley proofs were returned with the request to give us page proofs before 4 PM that day and we waited there. The page proofs were then given to us a little later and after correcting them we returned them the same day".

"Orders were placed for five hundred reprints of the bibliography. These were supplied within three or four days and 250 copies .... were distributed among the leading physicists and chemists of the world".

**

Note -7
{ref. 129}

"According to Krishnan's family sources, the diary was claimed by Ramanathan a day or two after Krishnan's death and it was returned to the family many years later, Chandrasekhar found it with Ramanathan in 1971. ......after finding that Krishnan did not leave the diary with the Royal Society, Chandra made a copy and deposited it with his own papers with the Royal Society Archives with his own recollections. These papers are currently inaccessible due to time embargo placed on them by Chandra himself. When the diary came back to Krishnan's family, it was found to have a total of only 16 pages covering the period of 5 to 28 February 1928, with the entry on the last day ending abruptly at an unfinished sentence that had started to describe the actual discovery. A substantial part of the diary appeared to be missing including the record of events of February 28 and all the subsequent work up to April 1928 that Raman and Krishnan did together to firmly establish the physical nature of the new phenomenon." It is clear that some vested interest did not want the truth of Raman-Krishnan controversy to come to light".

**

Note -8
{ref 131}

"The diary entry on the 7th February gave the first definitive clue:...Incidentally discovered that all pure liquids show a fairly intense fluorescence in the visible region, and what is much more interesting all of them are strongly polarized; polarization being the greater for the aliphates than for the aromatics. Infect the polarization of the fluorescent light seems in general to run parallel with the polarization of the scattered light, i.e. the polarization of the fluorescence light is greater the smaller the optical anisotropy of the molecule. When I told Professor about the results, he wouldn't believe that all liquids show polarized fluorescence and that in the visible region. When he came into the room, I had a bulb of pentane in the tank blue with violet filter in the path of incident light, and when he observed the track with a combination of green and yellow filters he remarked "you don't mean to suggest, Krisnayenger all that is fluorescence?!" However, when he transferred the green yellow combination also to the path of the incident light, he couldn't detect a trace of the track. He was very much excited and repeated several times it was an amazing result. One after another the whole series of liquids were examined and every one of them showed the phenomenon without exception. He wondered how we missed discovering all that five years ago......... After meals at night Venkateswaran and myself were chatting together in our room when Professor suddenly came to the house (at about 9 P. M.) and called for me. When we went down we found he was very much excited and had come to tell me that what we had observed this morning must be the Kramers-Heisenberg effect we had been looking for all these days. We therefore agreed to call the effect modified scattering rather than fluorescence".

Note -9
{ref 132}

Mallik and Chatterjee further write

"During the next two days Krishnan examined the new scattering phenomenon in a number of liquids. In the afternoon of the 9th, Krishnan had ether vapor as the sample and found a very pronounced signal of the new radiation. Raman was teaching in Science College and when he returned, Krishnan gave him a visual demonstration of the 'modified scattering' using sunlight as the source. Raman was ecstatic. As Krishnan wrote in the diary;".

"He ran about the place shouting all the time that it was a first-rate discovery,...He asked me to call in everybody in the place to see the effect" and immediately arranged in a most dramatic manner, with the mechanics to make arrangements for examining vapors at high temperatures".

"Evening was busy preparing the hot bath and I didn't go out. When Prof. returned after his walk he told me that I ought to tackle big problems like that........".

"Told Mr. Venkatswaran about the discovery and was discussing the problem with us, in the course of which he said that the phenomenon should be called Raman-Krishnan-Effect. Mr. Venkatswaran or somebody will call it by that name....".

**

Note -10
{ref. 163}

"'In a free running interview with B. R. Nunda in 1967, Blackett selected his influence as military consultant as probably more important than his other roles as scientific inventor. This is in marked contrast to Indian perceptions of him, which focused mainly on his influence on large scientific research organizations.......... Nehru identified with Blackett because both had been to Cambridge, held favourable attitudes to "political" socialism, and were cautious about the same kind of people, including the ' the Americans'".

**

Note -11
{ref 168}

"From the late 1940s, the scientists Blackett knew were travelling regularly; Bhatnagar to Norway to negotiate a heavy water deal, Bhabha to Ottawa to negotiate a reactor, Kothari to Moscow to purchase troop transport aircraft, Mahalanobis to Washington to look at the new large computers. As they passed through London, they all kept in touch with Blackett. Blackett visited their institutes, gave lectures there, examined their doctoral students, helped select candidates for appointments, appraised new research programmes and then promoted them if he liked them. These same scientists were also friends with A.V Hill, whose influence on science and the military in India began in 1943, well before Blackett's own relationships with that country. The Blackett friendship in some senses was an extension of the Hill friendship in the professional sense of advocacy within the scientific community. Hill knew Bhatnagar very well, and he also advised Bhabha on the activities of Bhabha's own institute. As personal friends of Blackett's, these men also asked him to watch out for their children and other relatives when they studied or worked in London, which he did."

**

Note -12
{ref 170}

"Daulat Singh Kothari, whom he helped to become the Scientific Adviser to the Minister of Defence and who thus headed the new Defence Science Organization, had first met Blackett in the Cavendish Laboratories in Cambridge in the early 1930s. Trained by Saha at Allahabad, Kothari was a theoretical astrophysicist who had seen how Rutherford's experimental laboratory was organized, and how stars in it like Blackett functioned. This new Defence Science Organization in Delhi was modelled on the one Blackett had just prescribed for the UK. Since it was first housed in the new National Physical Laboratory (of the CSIR), and borrowed scientists and equipment from it, there was around 1950 a deep integration of personnel in defence research and industrial research. The close relationship between Bhabha, Bhatnagar, Kothari and Blackett-and all of them with Nehru-reinforced the structural advances of such integration. Kothari now joined the group of science developers who had institutes to build and positions to fill: within

months of starting, Kothari received a letter from his revered teacher Meghnad Saha enquiring about a job in defence research for one of his sons. By 1951 Blackett was channelling all Indian requests for employment (on defence matters) directly to Kothari"

**

Note 13
{ref. 178}

"*Bhatnagar's somewhat dramatic rise to Secretary-level (thereby establishing a bench-mark for the scientists of the future) was in the Ministry of Education and at the instance of Maulana Abul Kalam Azad, the first Education Minister. He succeeded Sir John Sargent as Educational Adviser (and Education Secretary) to the Government of India in late 1947. In this context, the following entry of 21st December 1947 in the diaries of Alan Campbell-Johnson, Press Attaché to the last Viceroy, Lord Mountbatten, is of interest:.... Sri Krishna, one of the best informed of the Delhi political correspondents, referred to reports of a split in the Cabinet and claimed that the immediate cause of tension between Nehru and Patel was the action of Maulana Azad......Patel has recently set-up a sub-committee of top officials......to wet the appointments of all higher-grade civil servants. Azad has just appointed the well-known scientist, Bhatnagar, who is not a career man, as the Principal Secretary to his Ministry, without reference to this sub-committee....".*

**

Note -14
{ref. 212}

Bhabha's letter to Saklatvala, March 12, 1944

My dear Sir Sorab,

I have for some time past nurtured the idea of founding a first-class school of research in the most advanced branches of physics in Bombay. I had intended putting my scheme before you in person on my next visit to Bombay, but as a result of a letter from Prof. Choksi I am now sending it in writing for your consideration, and I would be glad to have your views on it. If you so desire I am prepared to come to Bombay to explain the scheme to the Trustees in person.

The scheme I am submitting now is not one which has been hastily conceived. It has been germinating in my mind for nearly two years, and I recently discussed it at length with Prof. A.V. Hill both at Delhi and at Bombay. Prof. A. V. Hill , Senior Secretary of the Royal Society, apart from being in eminent scientist himself, is one who has a great and intimate knowledge of the organisation of science and scientific institutions in England, and the many valuable suggestions he made have been incorporated in the scheme as it stands now. The scheme has been set forth on the accompanying schedule and is a simple one, but I should like to make a few remarks to explain its background.

There is at the moment in India no big school of research in the fundamental problems of physics, both theoretical and experimental. These are however all over India competent workers who are not doing good work as they would do if brought together in one place under proper direction. It is absolutely in the interest of India to have a vigorous school of research in fundamental physics, for such a school forms the spearhead of research not only in less advanced branches of physics but also in problems of immediate practical application in industry. If much of the applied research done in India today is disappointing or of very inferior quality it is entirely due to the absence of a sufficient number of outstanding pure research workers who would set the standard of good research and act on the directing board in an advisory capacity. (For example, while the Department of Scientific and Industrial was founded in Great Britain in 1914, it was soon felt that it could not function properly without the appointment of an adequate advisory council for the organisation. The scientific Advisory Council was founded in Great Britain in 1915 and was consisted mainly of eminent scientists like Lord Rutherford, Sir W. L. Bragg, Lord Rayleigh, Sir James Jeana, Prof. A. V. Hill and others. Without the availability of a sufficient number of pure research workers of this standing to serve on the Advisory Council, the work of the department would have suffered, as in India). However, when nuclear energy had been successful applied for power production in say a couple of decades from now, India will not have to look abroad for its experts but will find them, ready at hand. I do not think that anyone acquainted with scientific development in other countries would deny the need in India for such a school as I propose.

The subject on which research and advanced teaching would be done would be theoretical physics, especially on fundamental problems and with special reference to cosmic rays and nuclear physics, and experimental research on cosmic rays since the two are closely connected theoretically.

For the location of the school I think Bombay would be the most suitable place in India for the following reasons. Firstly, it is an advantage for a cosmic ray laboratory to be situated near the sea, for it is often necessary to make measurements at considerable depth under water. Secondly, Bombay as one of the first and most progressive cities in India has not yet got the scientific research institution necessary for its population and worthy of its position. People in educational circles in Bombay have long felt and expressed the urgent need for a good school of physical research. Thirdly, I feel that once a laboratory like the one proposed is established in Bombay, it will be easier to collect further money for it in addition to what Tata Trust may give. I am confident that both the government and the university would be prepared to give regular financial support.

In connection with the third reason I may mention confidentially that the Director of Public Instruction has on several occasion asked me if I would accept a chair at the Royal Institute of Science if one were created there for me with especially favourable conditions for research. He, I am sure, would get the Government to help the scheme of the type I propose. The institute would be affiliated to the Bombay University. The Bombay University could also send its advanced research students to the laboratory for work on their doctorate theses, and for attending the few advanced courses of lectures that we would give.

I also hope that in time we shall receive liberal support from the Board of Scientific and Industrial Research whose avowed policy include support for pure research, as publicly stated by Sir Ramaswamy Mudaliar when he presided at a lecture given by me to Delhi University this January. It would neither be feasible nor advisable to try to do research such as I plan under the same roof as applied physical research and routine testing, and it would be in the interest of efficiency if the Board of Scientific and Industrial Research decided to subsidise us to carry on pure research which it is its intention to foster by paying us say ten percent of the annual expenditure it contemplates on the projected National Physical Laboratory. Prof. Hill, when he was in

Bombay, repeatedly stressed the fact that all research has in the beginning to be built round a suitable man, and at the present moment there is no one else in India to do the type of research proposed. The same principle has guided the financing of research in Germany. To quote the words of the Director of the Kaiser Wilhelm Society which runs many of the biggest research institutions all over Germany

"In order that its ideals may be fulfilled, it is necessary that the Society should keep an intelligent watch on the newer currents in scientific investigation and try to further its ideals by creating facilities for new lines of investigation and by getting the right man for them. The object has thus been expressed by the president, Adolf V. Harnack, 'The K. W. Society shall not first build an institution for research and then seek out the suitable man but shall first pick up an outstanding man, and then build an institute for him.'

Experience has often shown that it is rather useful not only to call an outstanding man to the headship of an institute, but also to a group of associated institutions at one place and under a loose federation." Prof. A. V. Hill expressed the same views and added that was exactly the way in which outstanding school of research and ??? built up in United Kingdom, as for example the celebrated schools of physics and ????? at Cambridge. He saw no reason why the same thing could not be done here.

Financial support from Government need not, however, entail Government control, for to quote Prof. Hill in his lecture to the Science Congress at Delhi " Many of these (independent scientific institutions in Great Britain) now-a -days are receiving substantial state support: but nearly always when this is done a buffer of some kind is interposed to prevent Government support from becoming Government control."(Hill's underlining).

It might at first sight be supposed that the absence of a good school of physical research in Bombay at the moment would make it an unsuitable place for the object I have in mind. This is not so. The best and the most promising students desirous of studying theoretical physics or cosmic rays who for last three years have been sent to me in Bangalore from all parts of India, would come to Bombay instead. I am convinced that within five years we could make Bombay the centre of fundamental physical research in India.

Lastly, I would like to add a few personal remarks. It was while I was on holiday in 1939 that the war broke out and stopped my return to my job in Cambridge. For some time after that, I had the idea that after the war I would accept a job in a good university in Europe of America, because universities like Cambridge or Princeton provide an atmosphere which no place in India provides at the moment. But in the last two years I have come more and more to the view that provided proper appreciation and financial support are forthcoming, it is one's duty to stay in one's own country and build up schools comparable with those that other countries are fortunate in possessing. In 1941, I was offered the Physics Chair at the University of Allahabad with especially favourable conditions, and in 1942 the Professorship at the Indian Association for the Cultivation of Science in Calcutta, but I refused both these because I was not convinced that they afforded me sufficient scope for ultimately building up an outstanding school of physics. The scheme I am now submitting to you is but an embryo from which I hope to build up in the course of time a school of physics comparable with the best anywhere. If Tatas would decide to sponsor an institute such as I propose through their Trusts I am sure that they would be taking the initiative in a move which will be supported soon from many directions and be of lasting benefit to India.

With kind regards,
Yours sincerely

**

Note -15
{ref.234}

"Another major proposal for a big investment in scientific research came in 1898 from J N Tata, a leading industrialist of Bombay. The offer was sensational and attracted a great deal of attention at the turn of the century. Tata proposed to create a trust for the management of his income out of which 30 lakhs was to be applied to the founding and maintenance of a research institute, and the balance, estimated to exceed 30 lakhs, was to be held by the trustees for the benefit of Tata's descendants in perpetuity. He wanted the government to grant a similar amount to ensure the success and smooth functioning of the proposed institute. In one stroke Tata gained immense popularity, perhaps more than what M. L. Sarkar could get through his IACS movement. He symbolised the rising aspirations of the Indian bourgeoise which of late become conscious of the technological bearing of chemical and physical research."

"This put the government in a dilemma. In public it appreciated Tata's generosity, but in private officials doubted both the necessity and result of so

much expenditure on research in a colony. Moreover, Tata had lumped the proposal with his family fortunes and had asked for a legislation settling the balance of his property upon his descendants in perpetuity. He wanted his family to be saved from the application of the General Statute Law against the perpetuities. The Home Secretary to GOI wrote: 'This would form a very inconvenient precedent. I believe that there are not a few of the nouveaux riches if India who would be liberal on these terms, but the Government of India cannot afford to purchase their "munificence" on these terms. The secretary of state considered the proposals as 'nothing more or less than a bribe to the Government of India in order to obtain exemption from the general law as regard particular family settlement'. He added: I cannot see that political assistance in times of difficulty could be expected from men like Mr Tata, and in my opinion, such a departure from the principles of the general Statute law as Mr Tata asks for could only be justified on the ground of political necessity. Sensing trouble, Tata thought to be more prudent to withdraw the family clause. This took the wind out of the opposition's sails. Even Curzon confessed:' This ready acquiescence in our views indicates a liberality and unselfishness on his (Tata's) part for which I was hardly prepared. But in the summer of 1900 Tata went to London and tried to woo the Secretary of State over the Viceroy's head. Later in a private letter to the governor of Bombay, Curzon ruefully recalled:

Having been repeatedly told by us that we could not connect the foundation of the Institute with the creation of private endowment for his family, he (Tata) went behind our back to the Secretary of State, who did not know what had passed here, and wrung from him a consent to the coupling of the two schemes. He then wrote to us and asked us how we proposed to carry out the Secretary of State's orders. The later was as angry as we were, and he advised me to drop the whole thing. I was most reluctant to do this....".

.... Tata was no less determined. Enough pressure had been built in his favour. The Indian press was unanimous in its support A negative answer would have placed the entire blame on Curzon. The government could hardly afford to let this happen, so Curzon now tried to clip the proposals and asked for the scheme to 'be reduced to more modest dimensions.' Finally, he agreed to grant a sum of 2000 pound a year, but not without his usual impetuosity: 'Tata entirely owes it to me that he gets anything; and if he is not wise enough to accept it, I am ready to drop the whole thing tomorrow'. Tata refused to lie

low even though this meant the non-realisation of the scheme in his lifetime. He died in May1904, a partly disheartened man.....

.......Tata himself was called 'a crafty old man'. His initial attempt to lump family fortunes with the proposed institute gave a lever to Curzon who promptly projected it as part of a merchant's innate selfishness. ..."

**

Note -16
{ref. 242}

"It quite likely that Blackett and Bhabha may have discussed matters before the formal meeting, but at the same time Blackett must have been under considerable pressure with the risk of giving politically unsound advice to Nehru. Nehru first questioned Blackett on the "internal policy" of organising atomic research in India. Blackett began; "Nuclear physics is now being done by means of (a) big machines and (b) cosmic rays. Big machines are not worth having unless you have first class engineers and people who have the necessary flair for doing this sort of work".

**

Note -17
{ref.245}

"A community of users was established for the first time within the TIFR but for the AECI with the arrival of the Phillips machine. They were planning experiments with the machine; but the accelerator builders were not involved. Nuclear physicists working with the Philips machine soon found themselves in an uncomfortable situation with respect to research priorities of the TIFR and its surrogate work for the AECI. In April 1955, Evani Kondaiah, who had been hired from Oliphant's Birmingham laboratories, wrote to Bhabha: "I understand that a series of experiments are planned for the 'Reactor work' and the Cascade Generator will not be available for the Fundamental Research '.... So unless action is taken in this direction, now itself, I am afraid, the 'Fundamental Work' will suffer." Raja Ramanna (who led the nuclear physics groups of the AECI) considered Kondaiah's appeal and concluded that a very

small generator could not be purchased: those that could be bought were bigger and AECI did not need another machine. Reactor work would get priority on the machine because "this instrument was bought by the department for such work".

**

Note -18
{ref. 266}

"As soon as Dr. Bhatnagar came, the Prime Minister told him to accompany me to Bombay the next day and hold a meeting of the Council of Tata Institute, at which my resignation would be accepted. On the plane Dr. Bhatnagar made many derogatory remarks about Bhabha and advised me to remain firm".

"At the Bombay House where the council was to consider my resignation, Sir Saklatvala was the Chairman. Sir Ardeshir Dalal, one-time minister in the Central Government, Dr. Bhabha, Dr. Bhatnagar, and Mr. Moose, Director of public Instruction for Bombay province were also present".

"The Chairman said, "Prof. Gill, we have decided to accept your resignation, and you should write a letter of apology to the Director of the Institute, Dr. Bhabha, for returning to Bombay before the expiration of your deputation abroad".

"Surprised, I remarked, "'Sir, on the contrary, the Institute should congratulate me for returning to work after successfully concluding my foreign assignment." I walked out of the room".

"Dr. Bhatnagar came out and said. "They are very powerful people. They will ruin you unless you write a letter of apology as directed by the Chairman".

"That same evening, we took the Frontier Mail to Delhi. At the Bombay Railway station, Mr. Godebole, Registrar of the Institute, met me and asked me to sign a letter he had prepared. I took it and wrote another letter to the effect that if my return from abroad earlier caused any inconvenience to the Director I was deeply sorry for it. Next day, I saw Pandit Jawaharlal Nehru and told him what had happened. ...".
**

Note -19
{ref. 269}

Dr. Hussain wrote to Bhatnagar:

"I am sorry you do not find it possible to assist us in setting up the Gulmarg Research Observatory which the University of Jammu and Kashmir and Aligarh Muslim university have agreed to establish. If you had just said you cannot help us. I would understand. You have, however, been good enough to give a number of reasons why the Central Government finds itself unable to help. I am afraid they leave me unconvinced.

I am aware of the financial position of the Government of India, but I am also aware that their interest in scientific research in this country is such that, with laudable courage, they have come up with money to scientific projects. Who could know it better than you? I know that the budget proposals have been framed, but does it imply that the Department of Scientific and Industrial research will not- for at least a year- consider grant of assistance for any scientific project that is not known to and approved by them today? I cannot easily believe it.

The proposal for supporting scientific work done at the universities elsewhere will, I am sure, come to you for consideration, and I feel sure they will be dealt with by you from funds placed at your disposal by the Parliament. I do not think you will go for supplementary grant from Parliament for every small amount you decide to grant to one project or other during the course of the year.

"Reason number two given by you seems to indicate that I had asked for funds to establish the Observatory. It is not so. The two universities have decided to establish the Research Observatory themselves. They have a house for it which, if you were to establish, would cost a good few lakhs.

"I wanted only some assistance to help us equip the place. You give such assistance even to individual universities, and do not, I am sure, insist on their making their laboratories administered by an inter-university organisation. In the case under consideration, there are at least two universities cooperating from the very start to establish a place of research.

Besides, it is our intention from the very beginning to make the Research Observatory a place of scientific cooperation, and we are planning to make special arrangements to accommodate scientific workers from other universities and institutes, as well as (those persons) at the Gulmarg Research Observatory. We hope the Observatory will grow into a real centre for scientific cooperation.

"I am grateful for pointing out to me that the site is not a suitable one. We have very competent advice to establish the Observatory where we are doing it, and in view of the facilities which the State of Jammu and Kashmir have been good enough to give us, we think any disadvantage could be reasonably ignored.

The Observatory will, of course, do the work that can be done there to advantage, and I trust there will be lots of it.

Besides the reasons you have been pleased to give in your letter, for not being able to help us, you were good enough to tell me when we met in Delhi, that the fact Observatory is to be situated in Kashmir makes it difficult for you to help, and that if it were established in cooperation with some other Indian University, you would have done your best to help. I must frankly tell you that I did not relish this sentiment. I regard Kashmir as much a part of India as any other. Our people do, our government does so. We cannot behave as if we were a sort of 'No Man's land.

As the Prime Minister very rightly pointed out only a couple of days ago, 'Under the Indian Constitution, the State of Jammu and Kashmir is an integral part of the Republic,' and it was an extraordinary argument that the territory in dispute should be considered as some kind of 'No Man's Land' until its future was decided.

Permit me to repeat that we do not want the Government of India to establish the Research Observatory at Gulmarg for us. We only want some help to be able to equip it. We shall do our best to see that the Observatory is used by as many scientific workers as possible; we feel happy that it has been rendered possible to establish the observatory by the cooperation of the two Indian Universities -not a common event in our university life.

I hope you will reconsider your decision and lend a helping hand to this promising scientific venture. You will be glad to know that, even if the Government of India cannot assist us at this stage, we shall try to go ahead with the work and deserve their attention and help by dint of the value of the work. An enlightened government will, I feel sure, not long refuse to acknowledge the right we hope to establish; the right of service".

**

Note -20
{ref. 307}

Jahanvi Phalkey has given an account in her thesis and it is reproduced here;

"After the second Atoms for Peace meeting in Geneva in September 1958, Bhabha sent K. A. George to Blackett's laboratory at the Imperial College, London for further training and research in fusion. George had earlier led the construction of the Van de Graaff machine at the TIFR. To begin with, Phadke, who remained in charge of the three groups, was not entirely convinced of the decision. He wrote to Bhabha about progress in plasma physics experiments and argued; "An extended stay of Mr. George at the Imperial College will retard our work considerably.... In my opinion, Mr. George is mature enough to benefit by a brief stay at various places. I therefore suggest the following programme for Mr George's deputation after the Geneva Conference: 3 months at the Imperial College, 3 days visit each at the following places: Stockholm, Uppsala, Aachen, Saclay and Munich". Bhabha did not agree with Phadke, George would benefit from a year's stay at Imperial. He would have to stay. In December that year, George wrote to Phadke that he wanted to come back to Bombay. "Getting to know the technique of the work here required only a short time.... I have already discussed with Prof. Blackett and with Dr. Latham, the leader of the High Temperature Group, that the main object of my stay was to learn the technique and to get in touch with the spirit of plasma research, so that it will be of help in starting a formal group there.... I feel that further stay is not going to contribute much towards that objective." George further argued that staying back was going to delay the work he had already started in Bombay and opportunities for original work would be lost. Especially given the shuffling of personnel in the TIFR groups (two from the group were moving on, one to the USA and another to new assignments within the AECI set up), George felt he had to be in Bombay for readjustments. Phadke agreed and forwarded the letter to Bhabha with disastrous results".

"Bhabha did not discuss George's return. He questioned instead the prerogatives of the group in proceeding with fusion research. "Any work on fusion must be a substantial project, if it is not to be relatively ineffective. No experimental work on fusion in the Institute should be started until written

orders from the Director have been obtained. It should be clear that I am not happy at the considerable number of important projects that you have in your charge, which have not come to a satisfactory and definite conclusion. Till some of the present projects at least are successfully concluded, I cannot agree to any new project being started under you." In a surprise move, Bhabha's note put the focus back on the machines earlier constructed by the accelerator groups, something that was never before discussed seriously in relation to the fundamental research at the TIFR. On December 29, 1958, a faculty meeting decided to appoint a committee to consider the "utilisation of accelerators being developed under Dr. D. Y. Phadke"".

**

Note -21
{ref. 311}

"On November 23, 1959, the accelerator group submitted a joint representation to Bhabha. They were responding, "to a faculty evaluation of their group" as having done no tangible work and the likely chance that they may not be able to pursue plasma physics work in the near future. They requested an inquiry into the work of the group and permission to continue research in plasma physics. Quite explicitly, they expressed concern about their future in the Institute. They began with an outline of work on the three machines evaluated late the previous year for utilisation by the atomic energy establishment. The group argued that work on the cyclotron was completed in a span of four years and made a request that their effort be compared with the Calcutta cyclotron. "We were made to understand by Dr. Phadke that this programme was undertaken so that a group of people may be trained for the construction of bigger accelerators. In those days, we were proud to hear remarks to this effect from the Director when he occasionally visited our group. No publication was intended on this small cyclotron.... Since there was no demand for the 12" cyclotron, it was dismantled in mid-1956 with the permission of Dr. Phadke so that the magnet could be used for plasma work." The plans for a proton-synchrotron, a 60-inch cyclotron, the Van de Graaff, and odd jobs were discussed leading up to the beginnings of plasma physics at TIFR. "When there was no future for the group, after the 60" cyclotron program was dropped. One got interested in fusion research after the First

Geneva Conference, and a report was produced in December 1955. This was done at the time when no published work was available."

"Reiterating Phadke and George, the note said, "It was the emphatically declared policy in the section that the work of the groups would not be assessed on the number of publications.... Since an institute like ours could have an accelerator programme, we believed until now that we just happened to do that work. Due to reasons beyond our control, the Institute did not undertake any major accelerator work. That is quite understandable". The death of accelerator building within the TIFR was well pronounced. The group had been in existence for seven years and wondered if their work was considered not useful, why they had not been informed for these years. There also appears to have appeared a rift between Phadke and the group. They questioned his knowledge of the faculty's opinions and why he had not bothered to inform them. Finally, they asked for that they be allowed to continue work on plasma physics because: "Most of us joined the Institute because we like academic life and research work. We just happened to be in the Instrumentation section"".

**

Note -22
{ref. 314}

"I do not read the papers still, but I have begun to take a slight interest in general politics in India. I agree with you in considering Democracy the only form of government which will give at least a tolerable rule, though it is very far from being an ideal. The Platonic ideal may be ruled out in the present day. You support it because it is 'the only possible one which is consistent with the freedom of man'. 'Liberty' writes Napoleon in a private letter 'liberty is only the need of the few the few whom nature has endowed with exceptional talents, and there is no danger in restricting it. The crowd loves equality.' I agree with him here. A rule by a very efficient man like him or an efficient aristocracy is very good from the point of view of culture at least, and most other points of view, but there is always the trouble of who is to succeed afterwards, and what would happen in case the man or aristocracy should degenerate. Thus, I vote for democracy. We have very little liberty. The greater the city in which one lives, the more restricted do one's liberties become. At Cambridge I so arrange most of my meals, that I do the maximum possible amount of work, and the

time of the meals vary from day to day, sometimes by two hours, because I have most of them alone. This is impossible when one is with friends or a family. Moreover, the love of the crowd for equality is only a form of envy. Napoleon gets over this by giving distinctions to those who are most intelligent and efficient because of this envy becomes intensified if it is for people who do not deserve their lot, as in the French revolution. I agree that for a really good and efficient Democracy there ought to be a widely diffuse culture through the masses. You will be surprised how uncultured the average Englishmen is. He has only acquired a certain sense of responsibility and a certain sense and forbearance in political matters through long experience in this form of government, which makes the government more or less stable. This sense can only be formed by experience, and therefore, it is unreasonable to expect a democracy in India which shall not make some blunders in the beginning, some even of a serious nature. We must remember France in the revolution. A man to have any self-respect, must have a certain background of culture, a certain pride in his race, which can only be cultivated by showing to him the greatness of his own race at various times in history, the high culture that they possessed, or how they advanced universal knowledge. India is not lacking in such past culture. It has one of the greatest in the world. What is required is to revive this ancient culture, to modernize it, to take and blend with it the greatest and best parts of European culture, for if a nation is to live, and lead the world, it must possess a culture which is alive, which continues to change and evolve. Our aim should be to produce a type which is capable of appreciating Eastern art as much as western, though he may have preference for one, not a type to whom anything Eastern or anything Western is incomprehensible. For that is the future of man. ………"

**

Note -23
{ref. 331}

"Raman wrote its preface. …..He specifically mentioned Krishnan's experimental work with uranium bombardment at the Cavendish and added that his attempts to establish nuclear physics research at the Institute since then had not materialised".

"...... proposals landed at the gates of the Atomic Research Committee where Bhabha was the chairman and Saha was a member.... In addition, Bhabha was a member of the Tata family, a patron of the Indian Institute of Science."

"Saha did not attend the meeting to decide on the proposals instead lobbied with Gyan Ghosh and Bhatnagar for support".

"The members on the Committee were Bhabha, Saha and K. S. Krishnan, all members of the AERC; H. J. Taylor, a physics teacher at the Wilson College, Bombay, who had recently begun collaborating with Bhabha in experimental cosmic ray physics research; and the director J. C. Ghosh and Registrar, A. G. Pai, of the Indian Institute of Science, Bangalore".

"Raman's plea of having funds from the Government of India for expansion of the department were brushed aside. ... Bhabha ...... outlined an argument that was to become national policy on nuclear research for the next two decades. He argued that since a large part of the Institute budget came from the Government of India, the Institute's own plans could not be drawn in isolation from the "wider policy of the government in scientific matters". This wider policy of the Government in "scientific matters" was read out to the meeting from the minutes of the AERC. The government wanted to centralise and concentrate. Quite obviously, Bhabha and Taylor did not manage to convince Raman. .... In June 1947, the decision to reject the Bangalore proposal all together was still not arrived at in the meeting ...".

**

Note -24
{ref. 332}

"Before the Committee could meet, in July 1947, Bhatnagar wrote a letter to Bhabha about "J. C. Ghosh's" plans to begin nuclear physics education and research in Bangalore, arguing that it had to be stopped. But clearly, Bhabha knew about this before. Bhatnagar had an eagle eye view over the organisation of science education and research.........The Institute, he felt, should not be allowed to establish a chair in nuclear physics for Krishnan because was in conflict with the development of Bhabha's laboratory – as the centre of nuclear research in India. Bhabha lost no time and sent a telegram of his agreement – further requesting Bhatnagar to write to Meghnad Saha *with this idea. At least at this moment, Bhabha appears convinced that Bhatnagar could convince Saha to vote negatively on the Bangalore proposal, but Bhatnagar was not thinking on the same lines. His letter to Bhabha the very next day suggested that both Ghosh's and Saha's move to establish nuclear physics laboratories should be suppressed through the representatives of the council. Whether Bhatnagar, or for that matter Bhabha, wrote to other council members of the IISc is not clear. Bhatnagar did write to R. Choksi, the Dorabji Tata Trusts representative on the council. Bhatnagar mentioned the primacy of the Tata Institute of Fundamental Research for nuclear physics research in India, and as such, the creation of another chair for Krishnan in nuclear physics would be an "unnecessary duplication" of Tata efforts and therefore unwise use of philanthropy. Bhatnagar enjoyed no mean support and influence with the Tata Trusts. He had secured funding for the establishment of three national laboratories of the CSIR from the Tatas in the recent years. His word was not to be taken lightly."*

**

Note- 25
{ref. 448, 449}

"*The role of universities in the promotion of scientific research cannot be taken for granted. The freedom of the universities is being increasingly threatened in many parts of the world, particularly in India*".

"*Simply put. It means that university laboratories will not be allowed to function except in a very restricted sphere. Can we accept this? The universities have to provide properly trained personnel for private industry, for the national laboratories, and the armed forces. You can well imagine the effect of poor universities on the future of the nation*".

"*By contrast, US university laboratories while doing restricted work for the Atomic Energy commission operate strictly under the professor in charge of the laboratory. All recognized workers are provided funds and facilities, with no interference, which is why most scientific workers remain in university laboratories even at lower remuneration. In fact, it is a problem inducing scientists to join government. Government scientists, in practice, are guided by their colleagues in the universities. As far as it is practicable funds are distributed to the laboratories based on nature and importance of the work going on*".

"In summation:

1   *The place of universities must in no way be subordinated to government institutions. At present, university graduates who would ordinarily prefer to stay in the university laboratories are attracted to government posts for reasons of power and money. In other countries, university scientists have power vested in them by tradition; i.e., industry and government approach universities for advice and guidance when the need arises. There is nothing more noble or honorable than to help mold the lives of young scholars for the betterment of the nation.*

2   *An atmosphere of pure research is possible only in the shelter of a university, and not under a bureaucracy where one has to follow rigid rules and red tape.*

3

University scientists should not be burdened with so much administrative work. Their teaching load should also be reduced considerably to enable them to devote their time more to guiding young workers. ..".

*

Index

Bibliography

1 Tomczak M , Science and technology in India Lecture 14 Science , Civilisation and Society:

https://www.mt-oceanography.info/science+society/lecture14.html

2 Excellent analysis can be found in a series of lecture Science , Civilisation and Society by Matthias Tomczak Flinders University of South Australia:

https://www.mt-oceanography.info/science+society/

3 Citation index data: https://www.scimagojr.com/countryrank.php

4 Visvanathan Shiv: The tragedy of K. S. Krishnan: A sociological fable: Current **Science,** Vol 75, No 11, 10 December 1998 pp 1272-1275

5 Ramanna Raja: Years of Pilgrimage: An Autobiography, South Asia Books (1991) ISBN-10: 067083792X, ISBN-13: 978-0670837922

6 Kakodkar A, Gangotra S: Fire and Fury: Transforming India's strategic identity, Rupa Publications India Pvt. Ltd, New Delhi (2019)

7 Rao C N R :Climbing the limitless ladder: A life in chemistry, Cambridge University Press India Pvt. Ltd. New Delhi, (2014) ISBN 9788175969124

8 Khushwant Singh: https://en.wikipedia.org/wiki/With_Malice_towards_One_and_All

9 History of education in the Indian subcontinent:

https://en.wikipedia.org/wiki/History_of_education_in_the_Indian_subcontinent

10 Ibid.,

11 Chakrabarti Pratik: Western Science in Modern India, Metropolitan Methods, Colonial Practices, i, Permanent Black, Ranikhet, (2010) pp. 34

12 Deepak Kumar: Science and the Raj, Oxford India (paper backs) (2006) , pp. 33

13 Chakrabarti Pratik: Western Science in Modern India, Metropolitan Methods, Colonial Practices, , Permanent Black, Ranikhet, (2010) pp. 35-41

14 Deepak Kumar: Science and the Raj, Oxford India (paper backs) (2006) pp 66

15 Ibid., pp. 57-64

16 Deepak Kumar: Ibid., pp. 88

17 Deepak Kumar Ibid., pp. 89

18 Gaikwad, K D: Poona agriculture College: Catering to the 'Colonial Food' Requirement. 1908-47; in Science and Modern History, c. 1784-1947' Ed. Uma Dasgupta; Vol XV Part 4.: History of Science, Philosophy and Culture in Indian Civilisation; Ged D P Chattopadhyaya, Pearson Longman Delhi (2011) pp. 321-6

19 Pandya, S: Medical Education in Western India; Grant Medical College and Sir Jamsetjee Jejeebhoy's Hospital, Cambridge Scholars Publishing, UK, (2019) pp. 289-300

20 Ramanna, M: The Haffkine Institute, 1899-1947 ; in Science and Modern History, c. 1784-1947' Ed. Uma Dasgupta; Vol XV Part 4.: History of Science, Philosophy and Culture in Indian Civilisation; Ged D P Chattopadhyaya, Pearson Longman Delhi (2011) pp. 563-590

21 Deepak Kumar Science and the Raj, Oxford India (paper backs) (2006) pp 107

22 Ibid., pp. 109

23 Ibid., pp. 111

24 Prateek Chakrabarti Western Science in Modern India, Permanent Black, (2010) pp. 105

25 Ibid., pp. 107

26 Ibid., pp 130-131

27 Ibid., (2010) pp. 300

28 Deepak Kumar: Science and the Raj, Oxford India (paper backs) (2006) pp. 56

29 Pratik Chakrabarti: Western Science in Modern India, Metropolitan Methods, Colonial Practices, Permanent Black, Ranikhet, (2010) pp. 55

30 Ibid., pp. 91

31 Jukka Jouhki: Orientalism and India, , ISBN 951-39-2554-4 / ISSN 1459-305X (2006)

32 Mahendralal Sircar: https://en.wikipedia.org/wiki/Mahendralal_Sarkar

33 Indian Association for Cultivation of Science :

https://en.wikipedia.org/wiki/Indian_Association_for_the_Cultivation_of_Science

34 Parameswaran Uma: C V Raman : A Biography, Penguin India (2011) pp. 48

35 Bhattacharyy Kankan: Sir Asutosh and Rise of Modern Science in India,

Indian Journal of History of Science, 50.3 (2015) pp. 420-428

36 Singh Jagjit: Abdus Salam - A Biography https://archive.org/stream/AbdusSalam-
ABiography-English-JagjitSingh/abdus-salam-jagjit_djvu.txt

37 Blanpied William A: "Pioneer Scientists in Pre-Independence India"

Physics Today 39, 5, 36 (1986) https://doi.org/10.1063/1.881025

38 Ibid.,

39 Ibid.,

40 Wali K C: Chandra A biography of S. Chandrashekhar:;

University of Chicago Press (1991) pp 246.

41 Basalla George: The Spread of Western Science in Sal. P. Restivo and

Christopher K. Vanderpool, eds., Comparative Studies in Science and Society;

Charles Merrill Publishing Co.: Columbus, Ohio, (1974) pp. 359-381

42 Bhaumik, R: The History of Colonial Science and Medicine in British India: Centre-
Periphery Perspective, Indian Journal of History of Science, 52.2 (2017) 174-183

43 Kochhar R K: Science in British India, Indian Journal of History of Science,34 (4),

(1999) pp. 317-346

44 Ibid., pp. 328

45 Ibid., pp. 335

46 Patra S K and Muchie M: Science in pre-independent India: a scientometric

perspective; Annals of Library and Information Studies Vol64, (June 2017) pp. 125-136.

47 Ibid., pp. 127.

48 T R Seshadri: https://en.wikipedia.org/wiki/T._R._Seshadri#cite_note-Seshadri_on_WorldCat-11

49 N R Dhar : http://insaindia.res.in/BM/BM14_8901.pdf

50 Sir C V Raman: https://en.wikipedia.org/wiki/C._V._Raman

51 Ibid., pp. 129.

52 History of Education in Indian Subcontinent:

https://en.wikipedia.org/wiki/History_of_education_in_the_Indian_subcontinent

53 Kochhar R: Seductive orientalism: English education and modern science in colonial India, Social Scientist, 36:, (2008) pp. 45-63

54 Kochhar R: Cultivation of Science in the 19th Century Bengal,

https://www.academia.edu/767228/Cultivation_of_Science_in_the_19th_Century_Bengal?email_work_card=view-paper

55 History of education in the Indian subcontinent:

https://en.wikipedia.org/wiki/History_of_education_in_the_Indian_subcontinent

56 Syed Mahmood : A History of English Education in India (1781-1893), The Honorary Secretary of M A O College, Aligarh (1895) pp. 47

57 Ibid., pp. 87

58 Ibid., pp. 97

59 Industrial revolution : https://en.wikipedia.org/wiki/Industrial_Revolution

60 Oldest universities in UK:

https://www.study.eu/article/these-are-the-10-oldest-universities-in-the-uk

61 Asiatic Society of Bengal: https://en.wikipedia.org/wiki/The_Asiatic_Society

62 Royal Society: https://en.wikipedia.org/wiki/Royal_Society

63 Govil G: History of Science, Philosophy and Culture in Indian Civilisation, G. Ed. D P Chattopadhyaya, Volume XV Part 4 Science and Modern India: An Institutional History, c 1784-1947, Ed. Uma Dasgupta (2011) pp. 143-156

64 Ibid., pp. 147,149,151

65 Anderson R S: Nucleus and Nation: Scientists, International Networks, and Power in India (2011) pp. 83,

66 Sur, A: Scientism and social justice: Meghnad Saha's critique of the state of science in India. Historical Studies in the Physical and Biological Sciences, 33 (1), (2002) pp.87-105. doi:10.1525/hsps.2002.33.1.87 and references therein pp. 87.

67 Jayram Ramesh: Nehru's scientific temper recalled, Text of 13th Convocation Address delivered at IIT, Guwahati, May 27th, (2011) pp. 3

68 Govil G: Indian science Congress and the three Academies, 1914-35, in History of Science, Philosophy and Culture in Indian Civilisation, G Ed. D P Chattopadhyaya, Volume XV Part 4 Science and Modern India: An Institutional History, c 1784-1947, Ed Uma Dasgupta (2011) pp. 146

69 Ibid., pp. 149

70 Indian Science Congress Association: https://en.wikipedia.org/wiki/Indian_Science_Congress_Association

71 INSA: http://www.insa.nic.in/UI/pagecontent.aspx?pc=OA=

72 Govil G. Indian science Congress and the three Academies, 1914-35 in History of Science, Philosophy and Culture in Indian Civilisation, G Ed. D P Chattopadhyaya, Vol XV Part 4 Science and Modern India: An Institutional History, Ed. Uma Dasgupta c 1781-1947. (2011) pp. 148

73 Ibid., pp. 148

74 Lewis Leigh Fermor: https://en.wikipedia.org/wiki/Lewis_Leigh_Fermor

75 Ramasesan S: The United Academy of Sciences of India-A piece of history, Current Science, Vol 67, No 9 & 10 pp. 633-635

76 Ibid.,

77 Ibid.,

78 Ibid.,

79 Ibid.,

80 Ibid.,

81 Ibid.,

82 Kocher R: Shanti Swarup Bhatnagar: Life and Times. Resonance (April 2002) pp. 82-89

83 Sur Abha: Scientism and social justice: Meghnad Saha's critique of the state of science in India, HSPS. Volume 33. Part I. ISSN 0890-9997, The Regents of the University of California, (2002) pp. 87-105

84 DeVorkin D H: Quantum Physics and the Stars (IV) Meghnad Saha's fate, JHA XXV, Science History Publication Ltd. Provided by the NASA Astrophysics Data System (1994) pp. 157

85 Sur Abha: Scientism and social justice: Meghnad Saha's critique of the state of science in India, HSPS. Volume 33. Part I. ISSN 0890-9997, The Regents of the University of California, (2002)] pp. 87-105

86 Ibid.,

87 Ibid.,

88 DeVorkin D H: Quantum Physics and the Stars (IV) Meghnad Saha's fate, JHA XXV, Science History Publication Ltd. Provided by the NASA Astrophysics Data System (1994) pp. 163

89 Uma Parameswaran: C. V. Raman: A Biography, Penguin India (2011) pp 1-5

90 Ibid., pp. 159

91 Anderson R S: Nucleus and Nation: Scientists, International Networks, and Power in India (2011) Pp. 125

92 Chaudhury Indra, Dasgupta Ananya: A Masterful Sprit: Homi J. Bhabha, Penguin Books, (2010) pp. 8

93 Anderson R S: Nucleus and Nation: Scientists, International Networks, and Power in India (2011) pp. 126

94 Sur Abha: Scientism and social justice: Meghnad Saha's critique of the state of science in India, HSPS. Volume 33. Part I. ISSN 0890-9997, The Regents of the University of California, (2002) pp. 87-105

95 Ibid., pp. 87-105

96 DeVorkin D H: Quantum Physics and the Stars (IV) Meghnad Saha's fate, JHA XXV, Science History Publication Ltd. Provided by the NASA Astrophysics Data System (1994) pp . 162 and ref there in

97 Ibid., pp. 162

98 Parameswaran Uma: C V Raman: A Biography, Penguin India (2011) pp. 145

99 DeVorkin D H: Quantum Physics and the Stars (IV) Meghnad Saha's fate, JHA XXV, Science History Publication Ltd. Provided by the NASA Astrophysics Data System (1994) pp 161 and ref. there in.

100 Ibid., pp. 162

101 Sur Abha: Scientism and social justice: Meghnad Saha's critique of the state of science in India, HSPS. Volume 33. Part I. ISSN 0890-9997, The Regents of the University of California, (2002) pp. 87-105

102 Anderson R S: Nucleus and Nation: Scientists, International Networks, and Power in India (2011) pp. 45

103 Ibid., pp. 38

104 DeVorkin D H: Quantum Physics and the Stars (IV) Meghnad Saha's fate, JHA XXV, Science History Publication Ltd. Provided by the NASA Astrophysics Data System (1994) pp. 164 and ref. there in

105 Parameswaran Uma: C. V. Raman: A Biography, Penguin India (2011) pp. 171

106 Ibid., pp. 173

107 Ibid., pp. 174

108 Mallik D C V, Chatterjee S: Krishnan Srinivasa Kariamanikkam: His life and Works , Universities Press, Hyderabad (2012) pp. 133-134

109 Ibid., pp.139

110 Ibid., pp. 150

111 Ibid., pp. 151

112 Singh Rajendra: Raman and the discovery of Raman effect, ,Phys. perspect. 4 (2002) pp. 399–420

113 Parameswaran Uma: C. V. Raman: A Biography, Penguin India (2011) pp. 142

114 Raman Research Institute: digital depository.

http://dspace.rri.res.in/browse?type=dateissued&sort_by=2&order=ASC&rpp=20&etal=-1&year=-1&month=-1&starts_with=1906

115 Wali K C : Chandra: A biography of S Chandrasekhar, The University of Chicago Press (1991) pp. 128-146

116 Ibid., pp. 249

117 Ibid., pp. 248

118 Wali K C: Chandra A biography of S. Chandrashekhar:; University of Chicago Press (1991) pp. 251

119 Ibid., pp. 252

120 Ibid., pp. 252

121 Mallik D C V and Chatterjee S: Kariamanikkam Srinivasa Krishnan: His Life and Works, , Universities Press (2012) pp. 85-93

122 Raman, C. V. and Krishnan, K. S., "A New Type of Secondary Radiation," Nature **121** (1928), pp. 501–502

123 Mallik, D C V and Chatterjee, S: Kariamanikkam Srinivasa Krishnan: His Life and Works, , Universities Press (2012) pp. 91

124 Ibid., pp. 92-93

125 Ibid., pp. 93

126 Ibid., pp. 94-95

127 Gunasekaran S and Arunachalam S: Impact factors of Indian open access journals rising, Current Science , Vol. 103, No. 7, (10 October 2012) pp. 757

https://repository.arizona.edu/bitstream/handle/10150/299579/0757.pdf?sequence=1&isAll owed=y

128 Mallik D C V and Chatterjee S: Kariamanikkam Srinivasa Krishnan: His Life and Works, , Universities Press (2012) pp. 87-88

129 Ibid., pp. 87

130 Wali K. C: Chandra, A Biography of S. Chandrasekhar, Viking, (1990) pp. 251-3

131 Ibid., pp. 88-89

132 Ibid. pp. 89

133 Ibid., pp. 89-90

134 Mallik D C V and Chatterjee S: Kariamanikkam Srinivasa Krishnan:
His Life and Works, , Universities Press (2012) pp. 94

135 Ibid., pp. 99

136 Ibid.,) pp.102

137 Ibid., pp. 102

138 Raman Sir C V: "The molecular scattering of light" Nobel Lecture, December 11, 1930 pp. 270 https://www.nobelprize.org/uploads/2018/06/raman-lecture.pdf

Raman Sir C V: – Nobel Lecture. NobelPrize.org. Nobel Media AB 2019. Mon. 1 Apr 2019. https://www.nobelprize.org/prizes/physics/1930/raman/lecture/

Elsevier Publishing Company, Amsterdam, 1965MLA style: "Sir Chandrasekhara Venkata Raman - Nobel Lecture: The Molecular Scattering of Light". Nobelprize.org. Nobel Media AB 2014. Web. 19 Jun 2018.

http://www.nobelprize.org/nobel_prizes/physics/laureates/1930/raman-lecture.html

139 Raman Sir C V: "The molecular scattering of light" Nobel Lecture, December 11, 1930 https://www.nobelprize.org/uploads/2018/06/raman-lecture.pdf pp 271

140 Ibid., pp 101

141 Wali K C: Chandra, A Biography of S. Chandrasekhar, Viking, pp251-3 (1990) pp. 251

142 Brand J C D: The discovery of the Raman effect, , Notes Rec.R.Soc.Lotid.43,(1989) pp. 1-23 https://royalsocietypublishing.org/doi/pdf/10.1098/rsnr.1989.0001

143 Ekalavya : https://en.wikipedia.org/wiki/Ekalavya

144 Krishna V V: Organization of Industrial Research: The early History of CSIR, 1934-47, History of Science Philosophy and Culture in Indian Civilisation, General Ed. D. P. Chattopadhyaya , Vol. XV Part 4 Science and Modern India: An Institutional History, c 1784-1947. Ed. Uma Das Gupta (2011) pp. 162.

145 Ibid., pp. 162

146 Ibid., pp. 164

147 Council for Scientific and Industrial Research: http://csirhrdg.res.in/cpyls.htm

148 Rahman A, Bhargava R N, Qureshi M A, Pruthi S: Science and Technology in India, Indian Council for Cultural Relations New Delhi (1973) pp. 13-23

149 SHENOY R P, Defence Research & Development Organisation (1958-82), DRDO monograph series. Ed. A L Moorthy ISBN 81-86514-15-5 Ed: Defence Research & Development Organisation Ministry of Defence New Delhi - 110 011 (2006)

150 Defence Research and Development Organisation: https://www.drdo.gov.in/drdo/English/index.jsp?pg=genesis.jsp

151 Shenoy R P: Defence Research & Development Organisation (1958-82), DRDO monograph series. Ed. A L Moorthy ISBN 81-86514-15-5 Ed: Defence Research & Development Organisation Ministry of Defence New Delhi - 110 011 (2006) pp. 13

152 Shenoy R P: Defence Research & Development Organisation (1958-82), DRDO monograph series. Ed. A L Moorthy ISBN 81-86514-15-5 Ed: Defence Research & Development Organisation Ministry of Defence New Delhi - 110 011 (2006) pp. 50

153 Anderson, R S: Nucleus and Nation:- Scientists, International Network, and Power in India:, The University of Chicago Press, First Indian Edition (2011) pp. 91

154 Das, A M and Das, A K: A. V. Hill: A Report to the Government of India on Scientific Research in India (1944). Annotated Edition; Edited by: - Information and Communication Society of India, 2016.

155 Deepak Kumar: Science and Society in Colonial India: Exploring an Agenda, Science Technology & Society 6: 2 (2001) pp. 389 https://doi.org/10.1177/097172180100600205

156 Patrick Blackett: https://en.wikipedia.org/wiki/Patrick_Blackett

157 J D Bernal: https://en.wikipedia.org/wiki/John_Desmond_Bernal

158 Patrick M S Blackett

https://www.nobelprize.org/prizes/physics/1948/blackett/biographical/

159 Blackett P M S: Science and Technology in an Unequal World, Jawahar Lal Nehru Memorial Lectures (1967-1972), Bhartiya Vidya Bhavan, Bombay (1973) pp. 2

160 Robert Anderson : Blackett in India: Thinking Strategically about New Conflicts, in Patrick Blackett - Sailor, Scientist and Socialist, Ed: Peter Hore, Taylor & Francis e-Library, (2005) pp. 217-266

161 Ibid., pp. 261

162 Robert S Anderson: Patrick Blackett in India, Military consultant and scientific intervener, 1947-72 Part One, Notes Rec. R. Soc. Lond. 53 (2), 253-273 (1999) and Two, Notes Rec. R. Soc. Lond. 53 (3), 345-350 (1999).

163 Ibid., pp. 254

164 Blackett P M S: Science and Technology in an Unequal world, , Jawahar Lal Nehru Memorial Lectures1967-1972,Bharatya Vidya Bhavan, Bombay, (1973), pp. 3

165 Anderson R S: Patrick Blackett in India: Military Consultant and Scientific Intervenor, 1947-72. Part One, Notes Rec. R. Soc. Lond. 53 (2), (1999) pp. 258

166 Ibid., pp 259

167 Ibid., pp 258

168 Ibid., pp. 258-9

169 Ibid., pp. 266.

170 Ibid., and reference there in pp. 267

171 Ibid., pp. 262 (and reference there in)

172 Anderson R S: Patrick Blackett in India: Military Consultant and Scientific Intervenor, 1947-72. Part Two, Notes Rec. R. Soc. Lond. 53 (2), (1999) pp. 347.

173 Ibid., pp. 348

174 Ibid., pp. 349

175 Sur Abha: Scientism and social justice: Meghnad Saha's critique of the state of science in India, HSPS. Volume 33. Part I. ISSN 0890-9997, The Regents of the University of California, (2002) pp 101

176 Bhatnagar, A S: Shanti Swarup Bhatnagar: His Life And Work, ISBN: 81-85322-49-X Panjab University, Chandigarh (2014) pp. 95

177 Ibid., pp. 125

178 Ibid., pp. ii

179 Anderson R. S.: Nucleus and Nation, The University of Chicago press (2011) pp. 114

177 Anderson R S: Patrick Blackett in India: Military Consultant and Scientific Intervenor, 1947-72. Part Two, Notes Rec. R. Soc. Lond. 53 (2), (1999) , pp. 115

181 The selected works of Mahatma Gandhi V 5: Hind Swaraj (1952) pp. 71 http://www.gandhiashramsevagram.org/voice-of-truth/gandhiji-on-means-and-ends.php

182 Remembering an Indian Prometheus: https://parsikhabar.net/individuals/remembering-an-indian-prometheus-homi-j-bhabha/18856/

183 Prometheus: https://en.wikipedia.org/wiki/Prometheus

184 Phalkey Jahnavi: Science, State Formation and Development: The Organisation of Nuclear Research in India 1938-1959, PhD thesis, Georgia Institute of Technology, Atlanta, USA, December 2007, pp. 23

185 Ibid., pp. 51

186 Sreekantan, B V: The Different Phases of Cosmic Ray Research in the 20th century:
Their Role in High Energy Physics and Astrophysics

https://www.google.com/url?sa=t&rct=j&q=&esrc=s&source=web&cd=1&ved=2ahUKEw
jD46ve--_gAhWYWX0KHb6-C_cQFjAAegQIARAC&url=http%3A%2F%2Fgrapes-
3.tifr.res.in%2Findico%2FmaterialDisplay.py%3FcontrIbid.,%3D25%26materialId%3Dsli
des%26confId%3D271&usg=AOvVaw3u_rhDObAimZ-zgUK75QQD

187 Phalkey Jahnavi: Not Only Smashing Atoms: Meghnad Saha and Nuclear Physics in
Calcutta, 1938-48. History of Science Philosophy and Culture in Indian Civilisation
General Ed. D. P. Chattopadhyaya , Vol. XV Part 4 Science and Modern India: An
Industrial History, c 1784-1947. Ed. Uma Das Gupta (2011) pp. 1057-94

188 Chaudhury I and Dasgupta A :A Masterful Sprit: Homi J. Bhabha, Penguin Books,
(2010) pp. 52 .

189 Menon M G K: JRD Tata's Legacy : The Development of India through Science, Dr.
Vikram Sarabhai Distinguished Professor of Indian Space Research Organization

http://www.tifr.res.in/~alumni/Jrd-3.pdf

190 Phalkey Jahnavi: Science, State Formation and Development: The Organisation of
Nuclear Research in India 1938-1959, PhD thesis, Georgia Institute of Technology,
Atlanta, USA, December 2007. pp. 160

191 Ibid., pp. 160

192   Chaudhury I, Dasgupta A: A Masterful Sprit: Homi J. Bhabha, Penguin Books,
(2010) pp. 79

193 Ibid., pp. 54

194 Ibid., pp. 112

195 Zia Mian: in The Nuclear Debate: Ironies and Immoralities , Zia Mian and Ashish
Nundy, Fellowship in South Asian Alternatives Regional Centre for Strategic Studies
Colombo, Sri Lanka July 1998 ISBN 9558051039 pp. 1

http://www.princeton.edu/sgs/faculty-staff/zia-mian/Homi-Bhabha-Killed-A-Crow.pdf

196 Ibid., pp. 1

197 "Letter from Homi Bhabha to Sir Dorab of Tata Trust," March 12, 1944, History and Public Policy Program Digital Archive, Institute for Defence Studies and Analyses (IDSA), Tata Institute of Fundamental Research, Homi Bhabha Papers, IDSA-HBP-12031944. Wilson Center, Digital Archive; Obtained and contributed by A. Vinod Kumar and the Institute for Defence Studies and Analyses.

https://digitalarchives.wilsoncenter.org/document/114188

198 Chaudhury I, Dasgupta A: A Masterful Sprit: Homi J. Bhabha, Penguin Books, (2010) pp. 61 .

199 Chaudhury I, Dasgupta A: A Masterful Sprit: Homi J. Bhabha, Penguin Books, (2010) Letter from Rus to Homi dt. 17-02-1944.   pp. 94

200 Anderson R S: Nucleus and Nation:- Scientists, International Network, and Power in India:, The University of Chicago Press, First Indian Edition (2011) pp. 85

201 DeVorkin D H: Quantum Physics and the Stars (IV) Meghnad Saha's fate, JHA XXV, Science History Publication Ltd. Provided by the NASA Astrophysics Data System (1994) pp. 173

202 Anderson R S: Nucleus and Nation:- Scientists, International Network, and Power in India:, The University of Chicago Press, First Indian Edition (2011) pp. 87

203 Ibid., pp. 98

204 Ibid., pp. 104

205 Zia Mian: The Nuclear Debate: Ironies and Immoralities , Zia Mian and Ashish Nundy, Fellowship in South Asian Alternatives Regional Centre for Strategic Studies Colombo, Sri Lanka July 1998 ISBN 9558051039 pp. 3

http://www.princeton.edu/sgs/faculty-staff/zia-mian/Homi-Bhabha-Killed-A-Crow.pdf

206 Chaudhury I, Dasgupta A: A Masterful Sprit: Homi J. Bhabha, Penguin Books, (2010) pp. 94

207 Ibid., pp. 94

208 Ibid., pp. 94]

209 Ibid., pp. 94

210 Ibid., pp. 111

211 Ibid., pp. 95

212 "Letter from Homi Bhabha to Sir Dorab of Tata Trust," March 12, 1944, History and Public Policy Program Digital Archive, Institute for Defence Studies and Analyses (IDSA), Tata Institute of Fundamental Research, Homi Bhabha Papers, IDSA-HBP-12031944. Wilson Center, Digital Archive; Obtained and contributed by A. Vinod Kumar and the Institute for Defence Studies and Analyses. https://digitalarchives.wilsoncenter.org/document/114188

213 Virk, Hardev and Singh Rajinder:. The Pioneers of Cosmic Ray Research in India. Research and Reviews: Journal of Space Science and Technology 5 (2016) pp1-7.

214 B.V Srikantan: The Different Phases of Cosmic Ray Research in the 20th century: Their Role in High Energy Physics and Astrophysics, National Institute of Advanced Studies, IISC Campus, Bangalore 560012 https://grapes.3.tifr.res.in/indico/meteria……..

215 Roy S C and Singh Rajendra: D M Bose and Cosmic Ray Research,, Indian Journal of History of Science, 50.3(2015) pp. 438-455,

216 Singh Rajendra: D. M. Bose- His Scientific Work in International Context, Springer Verlag, (1916) ISBN 3844046199, 9783844046199,

217 Virk H. S. and Singh Rajendra : The Pioneers of Cosmic Ray Research in India, https://www.researchgate.net/publication/307926609_H._S._Virk_and_Rajendra_Singh Research and Reviews: Journal of Space Science and Technology ISSN:2321-2837 (online), ISSN:2321-6506 (print) Vol. 5 Issue 2, (2016)

218 Das M P: D. M. Bose: The Indian who missed the Nobel,, Science Reporter, November (2010), pp. 42-43

219 C F Powel: https://en.wikipedia.org/wiki/C._F._Powell

220 Powell C F: The study of Elementary Particles by the Photographic Method'
Pergamon Press, (1959), pp. 32

200 Roy S C and Singh Rajendra: D M Bose and Cosmic Ray Research, ,
Indian Journal of History of Science, 50.3 (2015) pp. 438-455

222 Singh Rajendra: D M Bose-His Scientific Work in International Context,
Rajender Singh, Shaker Verlag, (1916) ISBN 3844046199, 9783844046199

223 "Letter from Homi Bhabha to Sir Dorab of Tata Trust," March 12, 1944,
History and Public Policy Program Digital Archive, Institute for Defence Studies and
Analyses (IDSA), Tata Institute of Fundamental Research, Homi Bhabha Papers,
IDSA-HBP-12031944. [Wilson Center, Digital Archive; Obtained and contributed by
A. Vinod Kumar and the Institute for Defence Studies and Analyses]
https://digitalarchives.wilsoncenter.org/document/114188

224 Phalkey Jahnavi: Science, State Formation and Development: The Organisation
of Nuclear Research in India 1938-1959, PhD thesis, Georgia Institute of Technology,
Atlanta, USA, (December 2007), pp. 21.

225 Sreekantan, B V: The Different Phases of Cosmic Ray Research in
the 20th century: Their Role in High Energy Physics and Astrophysics
https://www.google.com/url?sa=t&rct=j&q=&esrc=s&source=web&cd=1&ved=2ahUKEw
jD46ve--_gAhWYWX0KHb6-C_cQFjAAegQIARAC&url=http%3A%2F%2Fgrapes-
3.tifr.res.in%2Findico%2FmaterialDisplay.py%3FcontrIbid.,%3D25%26materialId%3Dsli
des%26confId%3D271&usg=AOvVaw3u_rhDObAimZ-zgUK75QQD

226 Phalkey Jahnavi: Science, State Formation and Development: The Organisation of
Nuclear Research in India 1938-1959, PhD thesis, Georgia Institute of Technology,
Atlanta, USA, (December 2007), pp. 301

227 Ibid., pp. 188

228 Zia Mian: Homi Bhabha killed a crow: in The Nuclear Debate: Ironies and

Immoralities Ed. Zia Mian and Ashis Nandy Regional Centre for Strategic Studies
Colombo, Sri Lanka, (1998), pp. 4

229 Sharma, D: India's Nuclear Estate, Lancers Publisher, (1983) pp. 6

230 M G K. Menon: JRD Tata's Legacy : The Development of India through Science,
pp.12 http://www.tifr.res.in/~alumni/Jrd-3.pdf

231 Bhabha Homi: Note on the organisation of Atomic Research in India, April 26, 1948
in Homi Jahangir Bhabha on Indian Science and the Atomic Energy Programme: A
Selection, Tata Institute of Fundamental Research, (2009) pp. 67-72

232 Sharma, D : India's Nuclear Estate, Lancers Publisher, (1983) pp. 87

233 Kochhar R: Shanti Swarup Bhatnagar: Life and times
https://rajeshkochhar.com/data/publications/ssb_by_rkk.pdf

234 Kumar D: Science and the Raj, Oxford India paperbacks, (2006) pp.202

235 Phalkey Jahnavi: Science, State Formation and Development: The Organisation of
Nuclear Research in India 1938-1959, PhD thesis, Georgia Institute of Technology,
Atlanta, USA, (December 2007) pp. 295

236 Bhabha, H: Speech delivered at the inauguration of the Tata Institute of
Fundamental Research: Homi Jehangir Bhabha on Indian Science and the
Atomic Energy Programme: A Selection TIFR (2009) pp. 75-78

237 Ibid., Pp. 299-300

238 Ibid., pp. 296

239 Phalkey Jahnavi: Science, State Formation and Development:
The Organisation of Nuclear Research in India 1938-1959, PhD thesis,
Georgia Institute of Technology, Atlanta, USA, (December 2007) pp. 295

240 Ibid., pp. 301-2

241 Ibid., pp. 304-5

242 Ibid., pp. 314 and the ref therein

243 Ibid., pp. 330

244 Ibid.,

245 Ibid., pp. 340 and the ref therein

246 Mark Oliphant: https://en.wikipedia.org/wiki/Mark_Oliphant

247 Phalkey Jahnavi: Science, State Formation and Development: The
Organisation of Nuclear Research in India 1938-1959, PhD thesis,
Georgia Institute of Technology, Atlanta, USA, (December 2007) Pp 343

248 Ibid., pp. 344

249 Ibid., pp. 353-54 and the ref therein

250 Ibid., pp. 311 and the ref therein

251 Puri, S P: Prof. Piara Singh Gill, Biog. Mems. Fell. INSA, N. Delhi, 25, (2004)
pp. 139- 149, http://www.insaindia.res.in/BM/BM25_0411.pdf

252 D D Kosambi: https://en.wikipedia.org/wiki/Damodar_Dharmananda_Kosambi

253 K S Chandrasekharan: https://en.wikipedia.org/wiki/K._S._Chandrasekharan

254 E C G Sudarshan: https://en.wikipedia.org/wiki/E._C._George_Sudarshan

255 Gill P S: Up Against Odds: Autobiography of an Indian Scientist,
Allied Publishers Ltd. ISBN 817023364X(1992)

256 Ibid., pp. 85

257 Ibid., pp. 85

258 Ibid., pp. 85-86

259 Ibid., pp. 85

260 Ibid., pp. 86

261 Ibid., pp. 86-87

262 Ibid., pp. 90

263 Ibid., pp. 91

264 Ibid., pp. 93

265 Chowdhury I, Dasgupta A: A Masterful Sprit: Homi J Bhabha,
Penguin Books India, ISBN 9780143066729 (2010) pp. 92

266 Gill P S: Up Against Odds: Autobiography of an Indian Scientist,

Allied Publishers Ltd. ISBN 817023364X(1992) pp. 95-96

267 Ibid., pp. 97

268 Ibid., pp. 106

269 Ibid., pp. 147

270 Ibid., pp. 151

271 M G K Menon : https://en.wikipedia.org/wiki/M._G._K._Menon

272 Roy S C, and Singh Rajinder: Bibha Chowdhuri – Her Cosmic Ray Studies

in Manchester, Indian Journal of History of Science, 53.3 (2018) pp. 356-373

273 Raju, C K: Kosambi the mathematician

http://ckraju.net/papers/Kosambi-the-mathematician.pdf

274 Deshmukh C: Dr . Damodar Dharmanand Kosambi-Life And Work,

Translated by Suman Oak, (1993) pp 51

https://archive.org/details/Dr.DamodarDharmanandKosambi-LifeAndWork/page/n1

275 Ibid., pp. 51

276 Chowdhury Indira: Fundamental research, Self-reliance and Internationalism:

The revolution of the Tata Institute of Fundamental Research, 1945-47 in History of

Science, Philosophy and Culture in Indian Civilisation, G Ed. D P Chattopadhyaya,

Volume XV Part 4 Science and Modern India: An Institutional History, c 1784-1947,

Ed. Uma Dasgupta (2011) pp. 1107

277 Deshmukh C: Dr . Damodar Dharmanand Kosambi-Life And Work,

Translated by Suman Oak, (1993) pp. 55

https://archive.org/details/Dr.DamodarDharmanandKosambi-LifeAndWork/page/n1

278 Ibid., pp. 56

279 Ibid., pp. 57

280 Ibid., pp 53

281 Ibid., pp. 57

282 Ibid., pp. 80

283 Ibid., pp. 58

284 Ibid., pp. 69

285 Ibid., pp. 81

286 Ibid., pp. 81

287 Ibid., pp. 81

288 Ibid., pp. 82

289 Ibid., Pp. 83

290 Ibid, pp 82

291 Raju, C K: Kosambi the mathematician

http://ckraju.net/papers/Kosambi-the-mathematician.pdf

292 Dani, S G: Kosambi, the Mathematician, Resonance (June 2011) pp 514-528

https://www.ias.ac.in/article/fulltext/reso/016/06/0514-0528

293 Raju C K: Kosambi the mathematician

http://ckraju.net/papers/Kosambi-the-mathematician.pdf

294 Kienholz M: Opium Traders and Their Worlds-Volume Two: A Revisionist
Exposé of the World's Greatest Opium Traders, Chapter Thirty Three: - The Tata
Dynasty: Opium and Steel ISBN 9780595613267, iUniverse, (2008)

295 Deshmukh C: Dr . Damodar Dharmanand Kosambi-Life And Work,
Translated by Suman Oak, (1993) pp. 96

https://archive.org/details/Dr.DamodarDharmanandKosambi-LifeAndWork/page/n1

296 Raju, C K: Kosambi the mathematician

http://ckraju.net/papers/Kosambi-the-mathematician.pdf

297 Ibid., pp. 97

298 Raju C K: Kosambi the mathematician http://ckraju.net/papers/Kosambi-the-
mathematician.pdf

299 Seshadri, C S: K. Chandrasekharan (1920 – 2017) Bhāvanā, vol 1 i3   (July 2017) http://bhavana.org.in/k-chandrasekharan-1920-2017/

300 Sridharan, R: KC and I: My Fond Remembrance: Bhavana vol 2 i (January 2018) http://bhavana.org.in/kc-and-i-my-fond-remembrance/

301 Raju, C K: Kosambi the mathematician http://ckraju.net/papers/Kosambi-the-mathematician.pdf

302 Shaji, Anil: The Sciences (15/MAY/2018) https://thewire.in/the-sciences/remembering-e-c-g-sudarshan-a-seminal-theoretical-physicist

303 R E Marshak: https://en.wikipedia.org/wiki/Robert_Marshak

304 Luis J Boya: https://www.researchgate.net/profile/Luis_Boya

305 Marshak, R E: The Pain and Joy of a Major Scientific Discovery, Z. Naturforsch. 52a, pp .3-8 (1997)

306 Boya, L J: Laudatio for E. C. G. Sudarshan, Journal of Physics: Conference Series 87 (2007) pp. 1-6

307 Phalkey Jahnavi: Science, State Formation and Development: The Organisation of Nuclear Research in India 1938-1959, PhD thesis, Georgia Institute of Technology, Atlanta, USA, (December 2007) pp. 354-362 and reference there in

308 Ibid., pp. 358 and reference there in

309 Ibid., pp. 358-9 and reference there in

310 Ibid., pp. 359 and reference there in

311 Ibid., pp. 360-361 and reference there in

312 Shiraz Minwalla:
https://www.ictp.it/media/226739/ictp%20newsletter%20130_web.pdf

313 Hormasji Maneckji Seervai https://en.wikipedia.org/wiki/Hormasji_Maneckji_Seervai

314 Tewari Malvika: Homi to 'Homi'
https://issuu.com/malvikatewari/docs/book_homi_to_homi

315 Phalkey Jahnavi: Science, State Formation and Development: The Organisation of Nuclear Research in India 1938-1959, PhD thesis, Georgia Institute of Technology, Atlanta, USA, (December 2007) pp. 121

316 Banerjee H: Book Review Current Science ,

V 77, NO. 11, (10 December 1999) pp. 1548-15499.

317 Sundaram C V, Krishnan L V and Iyengar T. S.: Atomic Energy in India –

50 Years. Department of Atomic Energy, Government of India. (1998) pp. 277

318 Phalkey Jahnavi: Science, State Formation and Development: The Organisation of Nuclear Research in India 1938-1959, PhD thesis, Georgia Institute of Technology, Atlanta, USA, (December 2007) pp. 173

319 Ibid., pp. 58

320 Ibid., pp. 320

321 Ibid., pp. 279

322 Ibid., pp. 289

323 Ibid., pp. 279

324 Ibid., pp. 144

325 Ibid., pp.153

326 Ibid., pp.163

327 Ibid., pp. 164

328 Ibid., pp. 164,

329 Ibid., pp. 299-300

330 Ibid., pp. 164

331 Ibid., pp. 194

332 Ibid., pp. 189-190

333 Ibid., pp. 195

334 Ibid., pp. 196

335 Ibid., pp. 191-192

336 Banerjee H, Book Review Current Science, Vol. 77, NO. 11, 10 (December 1999) pp.1548-1549

337 Sundaram C V, Krishnan L V and Iyengar T S: Atomic Energy in India – 50 Years. Department of Atomic Energy, Government of India. 1998. pp. 277

338 Phalkey Jahnavi: Science, State Formation and Development: The Organisation of Nuclear Research in India 1938-1959, PhD thesis, Georgia Institute of Technology, Atlanta, USA, December 2007, pp. 122

339 Phalkey Jahnavi: Science, State Formation and Development: The Organisation of Nuclear Research in India 1938-1959, PhD thesis, Georgia Institute of Technology, Atlanta, USA, December 2007 "Chandrasekhara Venkata Raman (August 1947) C. V. Raman, Memorandum on Atomic Research in India, "Submitted to the Ministries of the Government of India" August 1947, RSK Papers, 3. pp. 139

340 Phalkey Jahnavi: Science, State Formation and Development: The Organisation of Nuclear Research in India 1938-1959, PhD thesis, Georgia Institute of Technology, Atlanta, USA, December 2007, pp. 346

341 Barman K C: Mahatma Gandhi's Morality: Ends and Means, EPRA International Journal of Multidisciplinary Research, Vol. 2,9 (September 2016) pp. 79-92

342 Anderson, R S: Nucleus and Nation:- Scientists, International Network, and Power in India:, The University of Chicago Press, First Indian Edition (2011) pp. 229

343 Deepak Kumar: Science and the Raj, Oxford India (paper backs) (2006) pp. 73

344 Anderson, R S: Nucleus and Nation:- Scientists, International Network, and Power in India:, The University of Chicago Press, First Indian Edition (2011) pp. 230

345 Mallik D C V, Chatterjee S: Krishnan Srinivasa Kariamanikkam: Universities Press, Hyderabad (2012) pp. 274

346 Bhatnagar S S: The National Physical Laboratory of India, Delhi, Current Science, Vol XVI (1947) pp. 71-73

347 Mallik D C V, Chatterjee S: Krishnan Srinivasa Kariamanikkam:, ,

Universities Press, Hyderabad (2012) pp. 325 and the ref therein

348 Kosambi D D: in D D Kosambi, Unsettling the past: ed Meera Kosambi, Permanent black (2014) pp. 88

349 Mallik D C V and Chatterjee S: Krishnan Srinivasa Kariamanikkam: His life and Works , Universities Press, Hyderabad (2012) pp. 352

350 Ibid., pp. 328

351 Science & Culture: http://www.scienceandculture-isna.org/journal.htm

352 Anderson R S: Nucleus and Nation:- Scientists, International Network, and Power in India:, The University of Chicago Press, First Indian Edition (2011) pp. 239-240

353 Venkataraman G.: Bhabha and his Magnificent Obsessions, University Press (1997) pp. 177-187.

354 Singh B B: Pullu The Nuclear Tadpole, Granth Ratnakar, Mumbai (2010) pp. 155-156

355 Needham Joseph: Science and civilization in China
https://archive.org/details/ScienceAndCivilisationInChina/page/n4

356 Chattopadhyaya D: Science and Society in Ancient India, Research India Publication, Calcutta, (1977)

357 Science, Civilisation and Society: Matthais Tomczak, https://www.mt-oceanography.info/science+society/lecture1.html

358 Science and Technology in India: Matthais Tomczak, http://www.mt-oceanography.info/science+society/lecture14.html

359 Science and Technology in China: Matthais Tomczak, http://www.mt-oceanography.info/science+society/lecture15.html

360 Science, technology and medicine in the Roman empire: http://www.mt-oceanography.info/science+society/lecture11.html

361 The rise of Hellenism. Alexandria, the new centre for science: http://www.mt-oceanography.info/science+society/lecture10.html

362 Greek Religion:

https://www.britannica.com/topic/Greek-religion

363 The new age of science in Greece: Matthais Tomczak,

http://www.mt-oceanography.info/science+society/lecture8.html

364 The rise of Hellenism. Alexandria: the new centre for science, Matthais Tomczak,

http://www.mt-oceanography.info/science+society/lecture10.html

365 Science, technology and medicine in the Roman Empire: Matthais Tomczak,

http://www.mt-oceanography.info/science+society/lecture11.html

366 Arrival of the Barbarians. The rise of Christianity: Matthais Tomczak

http://www.mt-oceanography.info/science+society/lecture12.html

367 Ibid.,

368 The state of science in medieval Europe: Matthais Tomczak

http://www.mt-oceanography.info/science+society/lecture13.html

369 Renaissance of the 12th century:

https://en.wikipedia.org/wiki/Renaissance_of_the_12th_century

370 Key Concepts of Puritanism and the Shaping of the American Cultural Identity: 1, Andreea MINGIUC, Philologica Jassyensia", An VI, Nr. 2 (12), (2010) pp. 211–217

371 Martin Luther: https://en.wikipedia.org/wiki/Martin_Luther

372 Revival of European Science, The New Cosmology Matthias Tomczak :

https://www.mt-oceanography.info/science+society/lecture19.html

373 Ibid.,

374 Protestant work ethics: https://en.wikipedia.org/wiki/Protestant_work_ethic

375 Puritans http://www.crf-usa.org/images/pdf/gates/puritans-of-mass.pdf

376 The English Revolution : https://oll.libertyfund.org/groups/68

377 Dechristianization of France during the French Revolution :

https://en.wikipedia.org/wiki/Dechristianization_of_France_during_the_French_Revolution

378 Age of Enlightenment : https://en.wikipedia.org/wiki/Age_of_Enlightenment

379 Concordat of 1801: https://en.wikipedia.org/wiki/Concordat_of_1801

380 Russian Revolution of 1917 : https://www.britannica.com/event/Russian-Revolution-of-1917

381 Religion in the Soviet Union :

https://en.wikipedia.org/wiki/Religion_in_the_Soviet_Union

382 https://www.nobeliefs.com/comments10.htm

383 Arabic and Islamic Science and Its Influence on the Western Scientific Tradition :

https://www.wdl.org/en/themes/islamic-science/timeline/#24

384 List of Muslim Scientists : https://en.wikipedia.org/wiki/List_of_Muslim_scientists

385 George A. Saliba: Science before Islam , in The Different Aspects of Islamic Culture,

Vol. IV, Science and Technology in Islam. Ed Raf: A. X al-Hmsan, at. al, UNESCO

Publishing (2001) pp 27

386 Indian influence on Islamic science :

https://en.wikipedia.org/wiki/Indian_influence_on_Islamic_science

387 Chinese influences on Islamic pottery :

https://en.wikipedia.org/wiki/Chinese_influences_on_Islamic_pottery

388 Greek contributions to Islamic world:

https://en.wikipedia.org/wiki/Greek_contributions_to_Islamic_world

389 Christian influences in Islam :

https://en.wikipedia.org/wiki/Christian_influences_in_Islam

390 Islamic Golden Age : https://en.wikipedia.org/wiki/Islamic_Golden_Age

391 Ibid.

392 Abbasid Caliphate : https://en.wikipedia.org/wiki/Abbasid_Caliphate

393 Imam Hamid el Gazali : https://en.wikipedia.org/wiki/Al-Ghazali

394 Taṣawwuf: https://en.wikipedia.org/wiki/Spirituality

395 Al-Ghazal: "Deliverance from Error", Translated, with related works, by Richard J.

Mccarthy, S.J., as Freedom and Fulfillment Boston, Twayne, C.S. 202

American University Of Beirut (1980)

396 Dr. Neil D Tyson lecture: https://www.youtube.com/watch?v=WZCuF733p88

397 Crusades: https://www.history.com/topics/middle-ages/crusades

398 https://www.thegreatcoursesdaily.com/the-mongol-sack-of-baghdad-in-1258/

399 https://www.thefamouspeople.com/17th-century-scientists.php

400 Wikipedia: https://www.wikipedia.org/

401 Vavilov S I: Soviet Science: Thirty Years, Marxists Internet Archive

https://www.marxists.org/archive/vavilov/1948/30-years/x01.htm

Archive https://www.marxists.org/archive/vavilov/1948/30-years/x01.htm

402 Loren Graham: Science in the New Russia, https://issues.org/graham-2/

403 Meiji Restoration: https://en.wikipedia.org/wiki/Meiji_Restoration

404 Meiji Restoration- Charter of Oath: https://en.wikipedia.org/wiki/Charter_Oath

405 Matthias Tomczak: Science and technology in China, http://www.mt-oceanography.info/science+society/lecture15.html

406 Long March and Cultural revolution in China:

https://en.wikipedia.org/wiki/Cultural_Revolution

407

https://www.indexmundi.com/facts/indicators/IP.PAT.RESD/compare#country=ca:cn:fr:de:in:it:jp:ru:gb:us

408 https://www.bostonglobe.com/ideas/2015/01/04/russian-science-amazing-why-hasn-taken-over-world/u61VuLiq3lJiyIMY0OLZ7N/story.html

409 Graham, Loren: "Science in the New Russia." Issues in Science and Technology 19, no. 4 (Summer 2003).

410 https://www.transparency.org/search

411

https://databank.worldbank.org/reports.aspx?source=2&series=IP.PAT.RESD&country=#

412 Canales M, McNaughton S: China's Science Boom, National Geographic

Vol 6 Issue 10 (2019) pp 25

413 Censorship, https://en.wikipedia.org/wiki/Censorship_by_country

414 Indian Scientists: https://en.wikipedia.org/wiki/List_of_Indian_scientists

415 Hindu Gurus and Saints:

https://en.wikipedia.org/wiki/List_of_Hindu_gurus_and_sants

416 Growth of India's population: http://statisticstimes.com/demographics/population-of-india.php

417 Kumbh Mela: https://en.wikipedia.org/wiki/Prayag_Kumbh_Mela

418 Acharya Rajneesh: https://en.wikipedia.org/wiki/Rajneesh

419 Maharshi Mahesh Yogi: https://en.wikipedia.org/wiki/Maharishi_Mahesh_Yogi

420 Kochhar R: J.C. Bose: The Inventor Who Wouldn't Patent, Science Reporter published by National Institute of Science Communication (NISCOM) "(February 2000)

421 Basalla G: The spread of western science, Science, Vol.156, (May 5, 1967) pp. 617

422 Ibid., 617

423 Kapila, Shruti: The Enchantment of Science in India, Isis, (2010) 101, 120–132 pp. 125 https://www.journals.uchicago.edu/doi/pdf/10.1086/652700

424 ISRO scientists performing pooja:

https://www.thehindu.com/2005/12/24/stories/2005122400250200.htm

425 Number 13 missing from PSLV series: https://www.dnaindia.com/technology/report-superstitions-and-beliefs-of-indian-space-scientists-1915176

426 Basalla G.: The Spread of Western Science, Science, Vol. 156 (May 5, 1967) pp. 617

427 Joseph Needham: The Grand Titration: Science and Society in East and West (London: Allen & Unwin, 1969)

428 Basalla G.: The Spread of Western Science, Science, Vol. 156 (May 5, 1967) pp. 618

429 Ibid., pp. 618

430 Ibid., pp. 618

431 Ibid., pp. 618

432 Ibid., pp. 618

433 Ibid., pp. 620

434 Trishanku : https://en.wikipedia.org/wiki/Trishanku

435 Deepak Kumar: The Trishanku Nation, Memory, self, and society in contemporary India, Oxford University Press, New Delhi (2016)

436 Baconian method: https://en.wikipedia.org/wiki/Baconian_method

437 S. Bhagavantam: http://www.indian-skeptic.org/html/is_v01/1-11-1.htm

438 Dr. Rajendra Prasad: https://en.wikipedia.org/wiki/Rajendra_Prasad

439 Deepak Kumar: The Trishanku Nation, Memory, self, and society in contemporary India, Oxford University Press, New Delhi (2016) pp. 148

440

https://www.msn.com/en-in/news/techandscience/newton-einstein-misled-the-world-kaurvas-were-test-tube-babies-speakers-at-indian-science-congress/ar-BBRPALl?ocid=spartanntp

441 Deepak Kumar: Emergence of 'Scientocracy': Snippets from Colonial India, Economic and political weekly 39 (35) (January 2004) pp. 3893-3898

442 Bhatnagar S S to A V Hill: 18 May 1951, A V Hill papers, AVHLII, Churchill College, Cambridge. Deepak Kumar, The Trishanku Nation, Memory, Self, and Society in Contemporary India, Oxford University Press, New Delhi (2016) pp.140

443 Anderson R S: Patrick Blackett in India: Military Consultant and Scientific Intervenor, 1947-72. Part Two, Notes Rec. R. Soc. Lond. 53 (2), (1999) pp. 348

444 Sir Ashutosh Mookherjee: https://en.wikipedia.org/wiki/Ashutosh_Mukherjee

445 United Kingdom Nobel Laureates:

https://en.wikipedia.org/wiki/List_of_Nobel_laureates_by_country#United_Kingdom

446 Sir Roland Ross: https://en.wikipedia.org/wiki/Ronald_Ross

447 Gill P S: Up Against Odds: Autobiography of an Indian Scientist, Allied Publishers Ltd. ISBN 817023364X(1992) pp. 83

448 Ibid., pp. 118

449 Ibid., pp. 120

450 Ibid., pp. 121

451 Tyabji Nasir: Jawahar Lal Nehru and Science and Technology, June 2007
https://www.researchgate.net/publication/257108160_Jawaharlal_Nehru_and_Science_an
d_Technology

452 Knuckle fight: https://mumbaimirror.indiatimes.com/mumbai/cover-story/senior-
scientists-in-a-bare-knuckle-fight-inside-barc/articleshow/56863934.cms

453 Vice-Chancellor sent to jail: https://timesofindia.indiatimes.com/india/Ex-VC-of-DU-
sent-to-jail-for-plagiarism-released/articleshow/45278628.cms

454 Vice Chancellors sacked: https://thewire.in/education/in-india-you-can-plagiarize-and-
flourish

455 Nobel Laurette to A P J Abdul Kalam : https://thewire.in/education/in-india-you-can-
plagiarize-and-flourish

456 IIT, Dhanbad plagiarism: https://www.thehindu.com/sci-tech/science/two-dozen-
papers-by-iit-dhanbad-faculty-retracted-so-far/article25681034.ece

457 Kochhar R, Falsifying scientific data: Blot on Indian science :
https://rajeshkochhar.com/tag/csir/

458 Scientific plagiarism in India:
https://en.wikipedia.org/wiki/Scientific_plagiarism_in_India

459 https://journosdiary.com/2019/07/20/authorships-73-papers-india-for-sale/

460 Rahman A, Bhargava R N, Quarishi M A, Pruthi S: Science and Technology in India,
Indian Council for Cultural Relations, New Delhi (1973)

461 Organizational chart of scientific and technological research, Ibid., pp. 49

462 Organizational chart of scientific and technological research, Ibid., pp 45

463 Organizational Chart ICAR, Ibid., pp. 52

464 Organizational Chart ,ICMR, Ibid., pp. 54

465 Organizational Chart, CSIR, Ibid., pp. 48

466 Organizational Chart, DRDO, Ibid., pp. 62

467 Organizational Chart Ibid.,, DAE, pp. 57

468 Organizational Chart ISRO:

https://www.isro.gov.in/about-isro/organisation-structure

469 Defence Research and Development Organisation:

https://www.drdo.gov.in/drdo/English/Defence_Research_and_Development_Organisation
.pdf

470 Department of Atomic Energy – Organizational Chart:

https://dae.nic.in/writereaddata/daeorg_eng.pdf

471 Bhabha Atomic Research Centre: Organizational Chart,

http://www.barc.gov.in/about/organ.html

472 Bhabha Atomic Research Centre: Employee List:

http://www.barc.gov.in/rti/emplist.html

473 Bhabha Atomic Research Centre: Delegation of powers BARC:

http://www.barc.gov.in/rti/sec_4_2.pdf

474 Bhabha Atomic Research Centre: Decision making process BARC:

http://barc.gov.in/rti/index.html#top

475 Total number of employees in all the purchase units of Directorate of Purchase
and Stores is about 1000: https://www.dpsdae.gov.in/english/Organisation.aspx

476 General financial rules 2017: http://www.iitk.ac.in/centralstores/data/GFR-2017-
Procurement.pdf

477 Manual for procurement of goods 2017:

https://doe.gov.in/sites/default/files/Manual%20for%20Procurement%20of%20Goods%20
2017_0_0.pdf

478 Rethinaraj T S G, Chakravarty Shoibal: Unethical Authorship is Research
Misconduct: https://sc-lab.org/pdf/2017-research-misconduct.pdf

479 Rethinaraj T S G, Chakravarty Shoibal: Unethical Authorship is Research Misconduct pp 20 https://sc-lab.org/pdf/2017-research-misconduct.pdf

480 Baldev Raj's rejoinder : https://www.thenewsminute.com/article/academic-wrote-77-papers-one-year-superhuman-brains-or-big-fat-lie-66524

481 Bare-knuckle fight -BARC: https://mumbaimirror.indiatimes.com/mumbai/cover-story/senior-scientists-in-a-bare-knuckle-fight-inside-barc/articleshow/56863934.cms

482 Prominent Physicist loses four more papers :
https://retractionwatch.com/2017/04/18/prominent-physicist-loses-four-papers-duplication/

483 https://retractionwatch.com/2017/05/12/top-physicist-loses-another-paper-duplication-tally-now-7/

484 BARC scientist who 'exposed plagiarism' faces harassment:
https://mumbaimirror.indiatimes.com/mumbai/crime/barc-scientist-who-exposed-plagiarism-faces-harassment/articleshow/54039742.cms

485 The Wire Report: https://thewire.in/uncategorised/sahoo-barc-plagiarism-physics

486 IMTECH scientist suspended : https://www.thehindu.com/news/national/CSIR-scientist-dismissed-for-fabricating-data/article14503466.ece

487 Jams Watson controversy: https://www.aljazeera.com/news/2019/01/lab-strips-dna-pioneer-james-watson-honours-racist-views-190114053532737.html

488 R P Shenoy: Defence Research and Development Organization (1958 to 1982), Defence Scientific Information and Documentation Centre, DRDO monograph series (2006) ISBN 81-86514-15-5

489 Phalkey Jahnavi: Science, State Formation and Development: The Organization of Nuclear Research in India 1938-1959, PhD thesis, Georgia Institute of Technology, Atlanta, USA, December 2007, pp. 343

490 Shenoy R P: Defence Research and Development Organization (1958 to 1982), Defence Scientific Information and Documentation Centre, DRDO monograph series, ISBN 81-86514-15-5 (2006) pp. 24

491 Gupta, A: T Seshadri (1900-1975) :

http://www.arvindguptatoys.com/arvindgupta/bs20trseshadri.pdf

492 N R Dhar: http://insaindia.res.in/BM/BM14_8901.pdf

493 G N Ramachandran: https://en.wikipedia.org/wiki/G._N._Ramachandran

494 Ramachandran plot: https://en.wikipedia.org/wiki/Ramachandran_plot

495 Olson, Richard G: Technology and Science in Ancient Civilizations

(Santa Barbara, CA: Praeger, (2010) , 4 , pp.113

http://www.questia.com/read/124045240/technology-and-science-in-ancient-civilizations

496 Said, Edward W: Orientalism: Western conceptions of the Orient.

Penguin Random House India (2001)

497 Manley, K: The Systems Approach to Innovation Studies, AJIS,

Vol 9 No-2 (May 2002) pp. 94-102

498 Godin, B Lane, Joseph P: "Pushes and Pulls": The Hi(story) of the Demand

and Pull Model of Innovation: Science Technology and Human Values,

Vol 38, Issue 5, (2013)

499 Ibid., pp. 33

500 Irrational comments at Indian Science congress:

https://scroll.in/article/908392/chauvanistic-claims-embarrassed-indian-scientists-protest-irrational-comments-at-science-congress

501 Joseph, M and Robinson, A : Free Indian Science, Nature,

V 508, 3 April 2014, pp. 35-38

502 Madan M, Gunasekaran S, Arunachalam S: Evolution of research in India:

Are we doing it right? Indian Journal of Medical Ethics, Vol III No 3 July-September

(1918) pp. 221-229

503 Deepak Kumar: The Trishanku Nation, Memory, self, and society in contemporary

India, Oxford University Press, New Delhi (2016) pp. 189

504 Sathyamurthy N, Bandyopadhyay R: Institution Building: The story of ISER's

https://www.ias.ac.in/public/Resources/Other_Publications/e-Publications/008/The_Story_of_IISERs.pdf

505 Indian Institute of Science Education and Research, Bhopal:

https://www.collegepravesh.com/research-colleges/iiser-bhopal/#ranking

**

www.ingramcontent.com/pod-product-compliance
Lightning Source LLC
Chambersburg PA
CBHW021358210526
45463CB00001B/150